Technological Spatula for Economic Sparkle

Alexis Jellybean

ISBN: 978-1-77961-754-5
Imprint: Preteen Chaos United
Copyright © 2024 Alexis Jellybean.
All Rights Reserved.

Contents

Introduction   1
The Importance of Technology in Economic Development   1
Historical Overview of Technological Transformations   35

Technological Infrastructure and Economic Development   63
Understanding the concept of technological infrastructure   63
Key Technologies for Economic Development   97

Technology Adoption and Diffusion   135
Understanding the process of technology adoption   135
Technology and Entrepreneurship   173

Technology, Innovation, and Economic Policy   207
Understanding the relationship between technology, innovation, and economic policy   207

Technology, Employment, and Inequality   243
Examining the impact of technology on employment   243

Technology and Globalization   271
Analyzing the impact of technology on globalization   271

Technology and Sustainable Development   303
Understanding the role of technology in sustainable development   303

Index   339

Introduction

The Importance of Technology in Economic Development

Understanding the role of technology in economic growth

In today's rapidly evolving world, technology plays a crucial role in driving economic growth. It is widely recognized that technological advancements have a significant impact on shaping the development trajectory of nations, industries, and individuals. In this section, we will explore the various dimensions of the role of technology in economic growth and understand its implications for societies.

The relationship between technology and economic growth

Technology and economic growth are closely intertwined. Technological advancements drive productivity gains, enhance innovation, and create new economic opportunities. When technology is effectively harnessed, it can lead to increased output, more efficient processes, and higher living standards.

To comprehend the relationship between technology and economic growth, let's consider the concept of the production function. The production function expresses the relationship between inputs (such as labor and capital) and outputs (goods and services). Technological progress is often considered as one of the factors that determine the total factor productivity (TFP), which captures the efficiency with which inputs are transformed into outputs.

$$Y = A \cdot F(K, L) \qquad (1)$$

In Equation (1), Y represents output, K denotes capital, L represents labor, and A represents a measure of technology or total factor productivity. Technological advancements, captured by the variable A, can lead to an increase in the efficiency of the production process, causing the production function to shift upward.

This upward shift in the production function signifies that with the same amount of inputs, greater output can be generated. As a result, per capita income increases, leading to economic growth. It is through technological progress that societies can achieve sustained increases in per capita income.

The impact of technology on productivity

One of the key mechanisms through which technology drives economic growth is by enhancing productivity. Productivity refers to the amount of output produced per unit of input. Technological advancements enable firms and individuals to achieve more with the same level of resources.

The use of advanced technologies, such as automation, artificial intelligence, and advanced analytics, can streamline production processes, reduce wastages, and eliminate inefficiencies. For example, automation technologies can replace manual tasks, leading to increased output per worker. Similarly, advanced analytics can provide insights to optimize resource allocation and decision-making, leading to improved productivity.

Moreover, technology can enable the creation of entirely new products, services, and industries, leading to increased economic activity. Innovations in sectors like information and communication technology (ICT), biotechnology, and renewable energy have led to the emergence of new markets and industries. These innovations have not only expanded employment opportunities but also increased overall productivity.

The relationship between technology and innovation

Innovation is a cornerstone of economic growth, and technology acts as a catalyst for innovation. Innovation refers to the process of translating new ideas, knowledge, or technologies into products, services, or processes that create value. Technological advancements often provide the raw materials for innovation, enabling the development of new solutions to existing problems or the creation of entirely new markets.

Technological innovations can take various forms, ranging from incremental improvements to disruptive breakthroughs. Incremental innovations involve small improvements in existing technologies, while disruptive innovations involve the introduction of entirely new technologies or business models that disrupt traditional industries.

THE IMPORTANCE OF TECHNOLOGY IN ECONOMIC DEVELOPMENT

The relationship between technology and innovation is a dynamic one. On one hand, technological advancements facilitate innovation by providing the tools and knowledge necessary for the development of new products and processes. On the other hand, innovation drives technological advancements by pushing the boundaries of what is possible and creating demand for new technologies.

The influence of technology on job creation

One of the significant concerns surrounding technological advancements is their impact on employment. While technology has the potential to automate certain tasks and replace human labor, its overall impact on job creation is more nuanced.

Technology not only replaces jobs but also creates new ones. As technologies advance, new industries, occupations, and job roles emerge. The introduction of automation technologies, for example, may lead to a decline in jobs that involve repetitive manual tasks but can also create a demand for jobs that require technical expertise to operate and maintain the automated systems.

Furthermore, technology can lead to the expansion of existing industries and the creation of new industries. It has the potential to drive economic growth and create employment opportunities through the development of new products, services, and markets. Consequently, the overall impact of technology on job creation depends on the interplay between automation and the emergence of new industries and occupations.

The connection between technology and income inequality

While technology has the potential to drive economic growth and create new opportunities, it can also exacerbate income inequality. The impact of technology on income distribution depends on various factors, including the sectoral composition of the economy, the skill requirements of new technologies, and the labor market dynamics.

Technological advancements often lead to a shift in the demand for different types of labor. Jobs that can be easily automated or outsourced may experience a decline in demand, leading to income losses for workers in those occupations. At the same time, jobs that require technical skills or creativity may experience an increase in demand, leading to higher incomes for workers in those occupations.

Furthermore, the ownership of technology and intellectual property rights can concentrate wealth and income in the hands of a few. As new technologies emerge, those who control and develop those technologies can accrue significant economic benefits.

To ensure that the benefits of technology are shared equitably, it is important to focus on policies that promote inclusive growth, improve access to education and training, and address skill mismatches in the labor market.

The role of technology in international trade

Technology plays a crucial role in driving international trade and reshaping global value chains. Technological advancements, such as telecommunications, transportation, and digital technologies, have significantly reduced the cost of communication and transportation, making it easier for countries to engage in trade.

Technological advancements enable firms to access new markets, connect with suppliers and customers globally, and engage in cross-border collaborations. For example, e-commerce platforms have facilitated international trade by providing a platform for small and medium-sized enterprises (SMEs) to reach customers worldwide.

Moreover, technology has enabled the fragmentation of production processes across different countries, leading to the emergence of global value chains. Global value chains refer to the sequential stages of production that take place in different countries. Each country specializes in specific stages of production, contributing to the overall value creation. Technologies such as just-in-time manufacturing, supply chain management systems, and digital platforms have enhanced the coordination and efficiency of global value chains.

The potential of technology in poverty reduction

Technology has the potential to contribute significantly to poverty reduction by creating new economic opportunities and improving access to basic services. Access to technology, especially digital technologies, can provide individuals with information, resources, and connectivity, enabling them to engage in economic activities and access markets.

For example, mobile phone technology has enabled individuals in rural areas to participate in mobile banking, access financial services, and engage in e-commerce. This has expanded economic opportunities for individuals who were previously excluded from formal financial systems.

Moreover, technology can improve access to essential services such as education, healthcare, and utilities. Digital technologies can facilitate remote learning, telemedicine, and the provision of clean energy in underserved areas, thereby reducing disparities in access to basic services.

However, it is important to ensure that technology is accessible and affordable for all segments of society. Efforts should be made to bridge the digital divide, improve digital literacy, and address the barriers that prevent marginalized communities from benefiting from technological advancements.

The challenges and risks associated with technology adoption

While technology has the potential to drive economic growth and improve quality of life, its adoption also presents challenges and risks. Some of the key challenges and risks associated with technology adoption include:

- **Cost and access barriers:** The cost of acquiring and implementing new technologies can be a significant barrier, especially for small and medium-sized enterprises (SMEs) and individuals in low-income countries. Limited access to technology can exacerbate existing inequalities.

- **Skills and education gaps:** The adoption of new technologies often requires a skilled workforce. The lack of necessary skills and education can hinder the effective adoption and utilization of technology.

- **Privacy and security concerns:** The increased reliance on technology raises concerns about data privacy and cybersecurity. The unauthorized use or disclosure of personal information can have significant social and economic implications.

- **Disruption and job displacement:** The rapid pace of technological advancements can disrupt industries and lead to job displacement. The transition to new technologies may be challenging for workers who need to adapt or acquire new skills.

- **Ethical considerations:** The development and use of technology raise ethical considerations, such as the impact on human rights, inequality, and privacy. It is important to address these ethical considerations to ensure that technology is developed and used responsibly.

Addressing these challenges and risks requires a comprehensive approach that involves policy interventions, investment in education and skills development, and the promotion of ethical guidelines for technology development and utilization.

Conclusion

In this section, we have explored the multifaceted role of technology in economic growth. Technology is a driver of productivity, innovation, job creation, and international trade. It has the potential to reduce poverty, improve access to services, and contribute to sustainable development. However, its adoption also presents challenges and risks that need to be addressed. By understanding the role of technology in economic growth and its implications, policymakers, businesses, and individuals can navigate the complex landscape of technological advancements and harness their full potential.

Discussing the impact of technology on productivity

In today's rapidly evolving world, technology plays a significant role in shaping economic development. One of the key aspects of this impact is its effect on productivity. Productivity, defined as the amount of output produced per unit of input (such as labor or capital), is a crucial determinant of economic growth and standards of living.

The relationship between technology and productivity

Technology has been a catalyst for improving productivity throughout history. By introducing innovative tools, machines, and processes, technology enables individuals and businesses to produce more goods and services with the same or even fewer resources. The positive relationship between technology and productivity is often referred to as the productivity paradox, where technological advancements lead to greater efficiency and output.

The impact of technology on productivity can be understood through various mechanisms:

1. **Automation and streamlining:** Technology enables the automation of repetitive and mundane tasks, freeing up human resources to focus on more productive and creative activities. For example, in manufacturing, the introduction of assembly line automation has significantly increased productivity by reducing production time and costs.

2. **Enhanced communication and collaboration:** The advancements in communication technology have revolutionized the way businesses operate. With the advent of email, video conferencing, and collaboration tools, employees can work together seamlessly regardless of their geographical

locations. This improves coordination, decision-making, and ultimately enhances productivity.

3. **Access to information and knowledge:** Technology provides easy access to vast amounts of information and knowledge resources. This enables individuals and organizations to make informed decisions, find innovative solutions, and stay up to date with the latest developments in their fields. Access to knowledge fuels innovation and improves productivity across sectors.

4. **Improved efficiency and accuracy:** Technology allows for the development of advanced tools and systems that increase efficiency and accuracy in various processes. For instance, computer-aided design (CAD) software helps engineers create precise and detailed designs, reducing errors and improving productivity in manufacturing and construction industries.

Challenges and risks associated with technology adoption

While technology has the potential to boost productivity, its adoption and implementation come with challenges and risks:

1. **Costs and investment:** Adopting new technologies often requires significant upfront investments in infrastructure, equipment, and training. For small businesses and developing economies, these costs can be a barrier to technology adoption, limiting their ability to enhance productivity.

2. **Skills and training:** Employing new technologies often necessitates acquiring new skills. The lack of proper training and skilled workforce can hinder the successful adoption of technology, undermining its potential impact on productivity.

3. **Unequal access and digital divide:** Socioeconomic disparities may result in unequal access to technology, creating a digital divide between different groups or regions. This divide can exacerbate existing productivity gaps, leading to increased inequality.

4. **Security and privacy concerns:** The increased reliance on digital technologies introduces new risks related to cybersecurity and data privacy. Security breaches and unauthorized access to sensitive information can have severe consequences, not only for individual businesses but also for the overall productivity and trust in the digital ecosystem.

Examples of technology-driven productivity improvements

Numerous sectors have experienced notable productivity improvements as a result of technological advancements. Here are a few examples:

1. **Manufacturing industry:** The introduction of robotics and automation technologies has transformed manufacturing processes, leading to significant productivity gains. Robots can perform repetitive tasks with high precision and speed, reducing production time and costs while improving product quality.

2. **Information technology sector:** The IT sector itself has significantly contributed to productivity improvements in various industries. The development of software applications, data analytics, and cloud computing has streamlined business operations, optimized resource allocation, and enhanced decision-making processes.

3. **Transportation and logistics:** Technological advancements in the transportation and logistics sector have revolutionized supply chain management, leading to increased efficiency, reduced transportation costs, and faster delivery times. Tracking systems, route optimization software, and automated warehouses have all contributed to these productivity improvements.

4. **Healthcare sector:** The implementation of electronic health records, telemedicine, and artificial intelligence-assisted diagnostics has enhanced productivity in the healthcare industry. These technologies have improved patient data management, facilitated remote consultations, and enabled more accurate and timely diagnoses.

Promoting productivity through technology adoption

To fully harness the potential of technology for productivity improvements, it is essential to address the challenges and promote widespread adoption. Here are some strategies:

1. **Investment in infrastructure:** Governments and organizations should prioritize investments in technological infrastructure, such as broadband connectivity and digital networks, to ensure equal access to technology across regions.

2. **Skill development and training:** Programs and initiatives that focus on equipping individuals with the necessary skills to adapt to technological changes are crucial. These efforts can include vocational training, reskilling, and upskilling programs to meet the demands of the evolving job market.

3. **Public-private partnerships:** Collaboration between government, private sector, and academia is vital for technology adoption. Public-private partnerships can facilitate the sharing of resources, expertise, and knowledge, fostering technology-driven productivity improvements.

4. **Support for innovation:** Governments can implement policies that encourage research and development, innovation, and entrepreneurship. Supporting innovation ecosystems can lead to the creation of new technologies and solutions that drive productivity growth.

5. **Addressing security and privacy concerns:** To instill confidence in technology adoption, robust cybersecurity measures and privacy regulations should be in place. Organizations should prioritize the protection of digital assets and data to ensure productive and secure technology utilization.

In conclusion, technology has a substantial impact on productivity by enabling automation, enhancing communication, improving access to information, and increasing efficiency. However, challenges such as costs, skill gaps, and unequal access must be addressed to ensure widespread adoption and maximize the productivity-enhancing potential of technology. Governments, organizations, and individuals all have a role to play in fostering a technology-friendly environment that promotes productivity and economic growth.

Examining the relationship between technology and innovation

Innovation is a key driver of economic growth and technological advancements play a crucial role in fostering innovation. Technology acts as a catalyst for innovation by providing new tools, resources, and possibilities for problem-solving and creating value. In this section, we will delve into the intricate relationship between technology and innovation, exploring how technological advancements fuel innovation and contribute to economic development.

Understanding the concept of innovation

Before we dive into the relationship between technology and innovation, it is important to have a clear understanding of what innovation actually entails.

Innovation can be defined as the process of developing and implementing new ideas, products, services, or processes that bring about positive change. It involves the creation, application, and diffusion of knowledge to generate economic and social value.

Innovation can take various forms, including incremental innovations that build upon existing technologies or processes, as well as disruptive innovations that fundamentally change the way things are done. It can occur in various sectors, ranging from technology and manufacturing to healthcare, agriculture, and finance.

Role of technology in fostering innovation

Technology serves as a critical enabler of innovation. It provides the foundation and tools for transforming ideas into practical solutions. Here, we will discuss the various ways in which technology fosters innovation:

1. **Enhancing research and development (R&D):** Technology plays a pivotal role in facilitating research and development activities. Cutting-edge technologies such as high-performance computing, advanced modeling and simulation, and data analytics enable scientists and researchers to conduct complex experiments and analyze large datasets. These tools help accelerate the pace of innovation by enabling faster experimentation, data-driven decision-making, and iterative improvements.

2. **Enabling new product development:** Technological advancements often open up new possibilities for product development. For example, the development of microprocessors and miniaturized electronics has paved the way for the creation of countless innovative consumer electronic devices such as smartphones, wearable devices, and smart home technologies. Additionally, advancements in materials science and manufacturing processes have enabled the development of new materials and products with enhanced properties and functionalities.

3. **Driving process innovation:** Technology also drives innovation in the realm of process improvement. Automation, robotics, and artificial intelligence have revolutionized manufacturing by enabling more efficient production processes and reducing costs. Similarly, advancements in information and communication technologies have streamlined business processes, enhanced supply chain management, and enabled the development of new business models.

4. **Facilitating knowledge dissemination and collaboration:** Technology has significantly improved the sharing and dissemination of knowledge, which is essential for fostering innovation. The internet, for example, has made it easier for researchers, entrepreneurs, and innovators to access information, collaborate with others, and share their ideas globally. Online platforms, open-source software, and crowdsourcing initiatives have democratized innovation by allowing a wider pool of individuals to contribute to problem-solving efforts.

Innovation-driven technological advancements

Innovation and technology are intricately intertwined in a continuous feedback loop. Technological advancements spur innovation, while innovation drives further technological developments. Here, we will explore some notable examples of innovation-driven technological advancements:

1. **Web and mobile technologies:** The advent of the internet and mobile devices has revolutionized our daily lives and the way businesses operate. The development of web and mobile technologies has given rise to a plethora of innovative applications and services, ranging from e-commerce platforms and social media networks to mobile banking and ride-sharing services. These innovations have transformed industries, created new business models, and disrupted traditional markets.

2. **Artificial intelligence (AI) and machine learning:** AI and machine learning have witnessed significant advancements in recent years, driving innovation in various sectors. AI-powered technologies, such as natural language processing, computer vision, and robotics, have enabled the development of intelligent chatbots, autonomous vehicles, medical diagnostic systems, and personalized recommendation algorithms. These innovations have the potential to improve efficiency, enhance decision-making, and transform industries such as healthcare, finance, and transportation.

3. **Renewable energy technologies:** Concerns about climate change and the need for sustainable development have fueled innovation in renewable energy technologies. Advancements in solar, wind, and hydroelectric power technologies have made clean energy sources more accessible and affordable. These innovations have not only contributed to reducing carbon emissions but have also created new opportunities for job creation and economic growth in the renewable energy sector.

4. **Biotechnology and pharmaceuticals:** The field of biotechnology has witnessed remarkable advancements, leading to significant innovation in healthcare and agriculture. Biotechnological innovations, such as gene editing, personalized medicine, and genetically modified crops, have the potential to revolutionize disease prevention, treatment, and food production. These technologies hold promise for improving health outcomes, enhancing crop yields, and addressing global challenges related to food security and environmental sustainability.

Challenges and risks associated with technology-driven innovation

While technology-driven innovation offers immense opportunities, it also poses certain challenges and risks that need to be addressed. Here are some key challenges and risks associated with technology-driven innovation:

1. **Ethical considerations:** Technological advancements often raise ethical concerns related to privacy, security, and equity. For example, the widespread use of artificial intelligence raises questions about the responsible and ethical use of data, the potential for algorithmic biases, and the impact on employment. Striking a balance between technological progress and ethical considerations is crucial to ensure that innovation benefits society as a whole.

2. **Technological inequality:** Access to innovative technologies is not evenly distributed, leading to technological inequalities. This digital divide can exacerbate existing social and economic disparities. Bridging the technological gap and promoting inclusive innovation is essential to ensure that the benefits of technology are shared equitably.

3. **Unintended consequences:** Technology-driven innovation can have unintended consequences that may be difficult to predict. For instance, the proliferation of social media platforms has raised concerns about the spread of misinformation, the impact on mental health, and the erosion of privacy. Anticipating and addressing these unintended consequences is crucial to ensure that innovation is socially responsible and aligned with the well-being of individuals and communities.

4. **Disruption of traditional industries and job displacement:** Rapid technological advancements can disrupt traditional industries and lead to job displacement. Automation and AI, for example, have the potential to

replace certain jobs and require individuals to acquire new skills. Navigating this transition and ensuring a smooth integration of technology into the workforce is essential for minimizing social and economic disruptions.

5. **Cybersecurity threats:** As technology advances, the risk of cybersecurity threats increases. Innovations in areas such as cloud computing, internet of things, and digital payments bring new vulnerabilities and challenges in protecting sensitive information. Developing robust cybersecurity measures and promoting cyber literacy are critical for safeguarding individuals, businesses, and critical infrastructure.

Promoting a culture of innovation

Creating an environment that fosters innovation requires a combination of factors, including supportive policies, investment in research and development, collaboration between academia and industry, and a culture that embraces and nurtures creativity and risk-taking. Here are some strategies for promoting a culture of innovation:

1. **Investing in education and research:** Building a strong foundation for innovation begins with investment in education and research. Providing quality education, fostering creativity and critical thinking skills, and promoting interdisciplinary research can spur innovation at an individual and organizational level.

2. **Encouraging collaboration and knowledge sharing:** Collaboration between academia, industry, and government is essential for driving innovation. Creating platforms and networks that facilitate knowledge sharing, collaboration, and open innovation can help generate new ideas and accelerate the innovation process.

3. **Supporting entrepreneurship and startups:** Startups and entrepreneurs often play a crucial role in driving technological innovation. Supporting entrepreneurship through access to funding, mentorship programs, and regulatory frameworks that encourage experimentation and risk-taking can create an ecosystem conducive to innovation.

4. **Promoting diversity and inclusion:** Embracing diversity and inclusion is crucial for driving innovation. Different perspectives and experiences foster creativity and help identify solutions to complex problems. Promoting diversity in the workforce and creating inclusive environments can unlock the full potential of innovation.

5. **Encouraging risk-taking and learning from failure:** Innovation is inherently associated with risks and uncertainties. Encouraging a culture that embraces risk-taking, tolerates failure, and encourages learning from failures can help foster a fearless attitude towards innovation.

In conclusion, the relationship between technology and innovation is a powerful force driving economic growth and societal progress. Technology serves as a catalyst for innovation by enabling new ideas, products, and processes. Advancements in technology fuel innovation, while innovation, in turn, drives further technological developments. However, the path towards innovation is not without challenges and risks, including ethical considerations, technological inequalities, unintended consequences, job displacement, and cybersecurity threats. By promoting a culture of innovation, investing in education and research, encouraging collaboration, supporting entrepreneurship, and embracing diversity, societies can unleash the full potential of technology-driven innovation and create a better future.

Exploring the influence of technology on job creation

Technology plays a significant role in shaping the labor market and has a substantial influence on job creation. As technology continues to advance, it leads to the emergence of new industries, changes in existing industries, and the creation of new job opportunities. In this section, we will explore the various ways in which technology affects job creation and the implications for individuals and the economy as a whole.

Understanding the relationship between technology and job creation

To understand the influence of technology on job creation, we need to first examine the relationship between the two. Technology has a dual impact on the labor market: it can lead to both job displacement and job creation. While technological advancements can automate certain tasks and replace workers in some industries, they also create new demands for labor in other industries and promote job growth.

Examining the impact of automation on job displacement

Automation, driven by technological advancements, has the potential to replace human workers in various sectors. Tasks that are repetitive, routine, and predictable are often the first to be automated. For example, in manufacturing industries, robots have replaced human workers in assembly line jobs. Similarly, in the retail sector, the introduction of self-checkout machines has reduced the need for cashier positions.

However, it is crucial to note that while automation may displace certain jobs, it does not necessarily lead to a decrease in overall employment. Instead, it shifts the composition of the labor market, creating a need for workers with different skills.

Understanding the role of technology in creating new job opportunities

While automation may replace certain jobs, it also creates new job opportunities. Technological advancements open up new industries and sectors, requiring workers with specialized skills. For example, the growing field of data analytics has created a demand for data scientists and analysts. Similarly, advancements in renewable energy technologies have led to the emergence of jobs in the green energy sector.

Technology also stimulates job creation indirectly by driving productivity gains and economic growth. As businesses become more efficient and productive through the adoption of technology, they often expand their operations and create additional jobs to support their growth.

Examining the skill-biased nature of technological advancements

Technological advancements tend to be skill-biased, meaning that they disproportionately benefit workers with higher levels of education and specialized skills. Jobs that require complex problem-solving, critical thinking, and creativity are less susceptible to automation and are more likely to be created as a result of technological progress.

On the other hand, routine and low-skilled jobs are more vulnerable to automation. Workers in these positions may face the risk of job displacement if they are unable to adapt and acquire new skills that align with the changing demands of the labor market.

Analyzing the need for upskilling and retraining

Given the skill-biased nature of technology, it is essential for individuals to continuously upskill and retrain to remain competitive in the job market. As technology evolves, the demand for certain skills changes, and workers need to adapt accordingly. Lifelong learning becomes crucial to ensure employability and job security.

Both individuals and governments have a role to play in facilitating upskilling and retraining initiatives. Individuals should proactively seek out opportunities to learn new skills, either through formal education or online courses. Governments can support these efforts by providing funding for training programs and implementing policies that promote continuous learning.

Discussions on the future of work

The influence of technology on job creation raises important questions about the future of work. As automation technologies continue to advance, there is an ongoing debate about the potential impacts on different job sectors and the overall structure of the labor market.

Some argue that technology will lead to widespread job losses and a significant disruption to the labor market. Others are more optimistic, suggesting that while certain jobs may be automated, new jobs will emerge, often in fields that we cannot yet foresee.

Additionally, the increasing use of artificial intelligence and machine learning raises concerns about the potential displacement of higher-skilled jobs that were historically considered safe from automation.

Exploring strategies for managing the impact of technology on job creation

To manage the impact of technology on job creation, various strategies can be implemented. These strategies include:

1. Promoting investment in education and skills development: By prioritizing education and training in areas relevant to emerging technologies, individuals can acquire the necessary skills to adapt to changing job requirements.

2. Encouraging a culture of lifelong learning: Individuals should be encouraged to pursue continuous learning throughout their careers to stay updated with technological advancements.

3. Facilitating transitions and retraining: Governments and industries should invest in programs that support worker transitions and provide opportunities for retraining in sectors with high job growth potential.

4. Fostering innovation and entrepreneurship: By supporting innovation and entrepreneurial activities, new industries and job opportunities can be created, leading to economic growth and job creation.

5. Collaboration between stakeholders: Governments, educational institutions, and industries should collaborate to identify skill gaps, design education programs, and create pathways for individuals to acquire the skills necessary for evolving job markets.

Real-world example: The impact of technology on the transportation industry

One real-world example of the influence of technology on job creation is the transportation industry. Technological advancements, such as autonomous vehicles and ride-hailing apps, have disrupted the traditional taxi industry but have also created new job opportunities.

While some traditional taxi drivers may have been displaced, the emergence of ride-hailing platforms like Uber and Lyft has created a demand for ride-share drivers. Additionally, the development of autonomous vehicles has opened up new possibilities for jobs in areas such as vehicle monitoring, maintenance, and software development.

This example illustrates how technology can both disrupt and create jobs within a single industry, highlighting the importance of adaptability and continuous learning for individuals working in sectors undergoing significant technological changes.

Further resources

To deepen your understanding of the influence of technology on job creation, the following resources may be helpful:

1. Book: "The Second Machine Age: Work, Progress, and Prosperity in a Time of Brilliant Technologies" by Erik Brynjolfsson and Andrew McAfee.

2. Article: "The impact of automation on employment: Just the usual structural change?" by Timo Boppart, et al. in the Journal of Monetary Economics.

3. Report: "The Future of Jobs Report" by the World Economic Forum, which explores the impact of technological advancements on job creation and identifies emerging job trends.

4. Website: edX.org offers a wide range of online courses on emerging technologies and their effects on the job market.

Conclusion

Technology not only disrupts and displaces jobs but also creates new job opportunities. The influence of technology on job creation is multifaceted, and its impacts vary across industries and skill levels. It is essential for individuals, industries, and governments to adapt and respond to technological advancements by fostering a culture of lifelong learning, upskilling, and implementing policies that support job transitions. By understanding the influence of technology on job creation, we can better prepare for the future of work and navigate the challenges and opportunities that lie ahead.

Analyzing the connection between technology and income inequality

Income inequality is a pressing issue that affects societies worldwide. As technology continues to advance at a rapid pace, it is crucial to analyze the relationship between technology and income inequality. In this section, we will explore how technological advancements can both exacerbate and mitigate income inequality, and discuss the various factors contributing to this connection.

Understanding the impact of technology on income distribution

Technology has the potential to impact income distribution in several ways. On one hand, technological progress can lead to increased productivity and economic growth, which in turn can create more job opportunities and higher wages, consequently reducing income inequality. For instance, the advent of computer technology has led to the automation of many repetitive and mundane tasks, allowing workers to focus on more complex and high-value work, leading to increased earnings.

On the other hand, technological advancements can also lead to job displacement and wage polarization. Automation and the use of artificial intelligence (AI) can replace routine jobs, especially in industries like manufacturing, leading to job losses and widening income disparities between skilled and unskilled workers. This phenomenon, known as technological unemployment, can exacerbate income inequality if proper measures are not taken to address it.

Examining the factors contributing to technology-induced income inequality

Several factors contribute to the connection between technology and income inequality:

Skills gap: Technological advancements often require workers with specialized skills in areas like AI, data analysis, and programming. Those with the necessary skills can benefit significantly from job opportunities provided by technological innovations, leading to increased income. However, workers without these skills may find themselves left behind, facing limited employment options and lower wages.

Education and training: Access to quality education and training programs is crucial in ensuring that individuals can acquire the skills needed to thrive in a technology-driven economy. Disparities in educational opportunities can contribute to income inequality, as individuals without access to education may struggle to take advantage of technological advancements.

Digital divide: The digital divide refers to the disparity in access to digital technologies, such as computers and the internet. Individuals or communities without access to these technologies face barriers in participating in the digital economy, which can limit their opportunities for income growth and exacerbate income inequality.

Concentration of wealth and power: Technological advancements have led to the rise of tech giants and the concentration of wealth in the hands of a few individuals or corporations. This concentration of wealth can contribute to income inequality as a significant share of the economic gains from technological progress is captured by a small elite.

Mitigating income inequality through technology

While technological advancements can contribute to income inequality, they also offer potential solutions to address this issue:

Educational reforms: Investing in education and training programs that equip individuals with the necessary skills to adapt to a technology-driven economy is crucial. Governments and educational institutions should collaborate to offer accessible and affordable training in emerging technologies, ensuring that no one is left behind.

Promoting digital inclusion: Bridging the digital divide through initiatives that provide affordable internet access and digital literacy programs can empower individuals with the tools they need to benefit from technology. This includes ensuring access to technology in underserved communities and providing training to enhance digital skills.

Supporting entrepreneurship: Encouraging entrepreneurship and innovation can create new avenues for income generation. Governments can provide support through funding, mentorship programs, and regulatory frameworks that foster innovation and entrepreneurship, particularly targeting underrepresented groups.

Real-world example: Microfinance and mobile technology

An excellent example of how technology can address income inequality is the combination of microfinance and mobile technology. In many developing countries, access to financial services is limited, leaving individuals without formal banking options.

With the widespread adoption of mobile technology, financial institutions have utilized mobile banking platforms to provide individuals with access to financial services, even those in remote areas. Microfinance initiatives have also leveraged mobile technology to offer small loans to micro-entrepreneurs, enabling them to grow their businesses and improve their incomes.

This integration of technology and finance has not only reduced income inequalities but has also empowered individuals economically, allowing them to overcome traditional barriers to financial services.

Conclusion

Technology plays a significant role in income inequality, both as a contributor and a potential solution. While technological advancements can widen income disparities, they also offer opportunities to mitigate inequality. By addressing the skills gap, promoting digital inclusion, and supporting entrepreneurship, societies can harness the potential of technology to create more equitable and inclusive economies. It is essential for policymakers, businesses, and individuals to collectively work towards harnessing the power of technology for the benefit of all, ensuring that no one is left behind in the digital age.

Assessing the role of technology in international trade

Technology plays a crucial role in shaping the landscape of international trade. In today's interconnected world, advancements in technology have revolutionized the way countries conduct business and engage in global economic activities. In this section, we will explore the various ways in which technology impacts international trade and discuss its implications for economic development.

Understanding the impact of technology on trade facilitation

Trade facilitation refers to the simplification, harmonization, and automation of trade processes, which can significantly reduce trade costs and improve efficiency. Technology plays a pivotal role in facilitating international trade by streamlining trade procedures and enhancing connectivity across borders.

One of the key contributions of technology to trade facilitation is through the automation of documentation processes. With the advent of electronic data interchange (EDI) and electronic documentation systems, paper-based transactions have greatly reduced, making trade transactions faster and more efficient. For example, digital platforms, such as blockchain technology, enable secure and transparent trade transactions, reducing the likelihood of fraud and errors. This promotes trust and confidence among trading partners, leading to smoother international trade operations.

Another significant impact of technology on trade facilitation is the development of online platforms and marketplaces. E-commerce platforms have transformed the way businesses engage in cross-border trade, enabling small and medium-sized enterprises (SMEs) to access global markets with ease. These platforms provide a virtual space for buyers and sellers to connect, eliminating the need for physical presence and reducing transaction costs. Technological advancements in logistics and transportation have also enabled faster and more reliable delivery of goods across borders, further enhancing trade facilitation.

Furthermore, technological innovations in customs administrations have led to more efficient border management processes. For instance, the implementation of automated customs clearance systems, such as single-window platforms, has streamlined the clearance procedures, reducing the time and cost of customs operations. Additionally, the use of advanced technologies, such as biometric identification and risk management systems, enhances border security while expediting the movement of goods across borders.

Analyzing the impact of technology on trade competitiveness

Technology has a profound impact on the competitiveness of nations in the global marketplace. Countries that are able to harness and adopt advanced technologies are better positioned to participate in and benefit from international trade.

One of the key ways technology enhances trade competitiveness is by improving productivity. By incorporating technology into production processes, businesses can achieve higher levels of efficiency, reduce costs, and produce higher-quality goods and services. This enables them to compete effectively in global markets. For example, the use of advanced manufacturing technologies, such as robotics and automation, allows for greater precision and speed in production, leading to increased competitiveness in industries ranging from automotive to electronics.

Moreover, technology enables companies to access and analyze vast amounts of data, leading to better market insights and improved decision-making. This allows businesses to identify emerging trends, target specific customer segments, and develop innovative products and services to meet evolving consumer demands. The ability to leverage technology for market intelligence gives companies a competitive edge in the international trade arena.

Technological advancements also facilitate access to global markets by reducing information asymmetry and transaction costs. Online platforms and digital marketing tools enable businesses to reach a wider audience, both domestically and internationally. This opens up new opportunities for SMEs and allows them to compete with larger firms in the global marketplace. Additionally, technology-enabled supply chain management systems enhance connectivity between suppliers, manufacturers, and distributors, leading to seamless coordination and improved competitiveness.

Understanding the challenges and risks of technology in international trade

While technology offers numerous opportunities for enhancing international trade, it also presents challenges and risks that need to be carefully managed.

One of the key challenges is the digital divide, which refers to the gap between countries and individuals with access to technology and those without. The uneven distribution of technology infrastructure and internet connectivity across countries can create disparities in trade opportunities. Developing countries with limited access to technology may face difficulties in participating fully in the digital economy and reaping the benefits of international trade. Addressing this digital divide through policies promoting digital inclusion and bridging the infrastructure gap is crucial for enabling equitable participation in international trade.

Another challenge is the rapid pace of technological change. Keeping up with technological advancements requires continuous investment in research and development, as well as the acquisition of new skills. Failure to adapt to new technologies can lead to a loss of competitiveness and marginalization in global markets. Governments and businesses must invest in education and training programs to ensure a skilled workforce capable of harnessing and leveraging technology for trade.

Furthermore, technology-driven globalization raises concerns about data privacy and cybersecurity. As businesses and governments increasingly rely on digital platforms for international trade, the protection of sensitive data becomes paramount. The risk of cyberattacks and data breaches highlights the importance of robust cybersecurity measures and international cooperation in establishing global standards for data protection.

In summary, technology has transformed international trade by enabling trade facilitation, enhancing trade competitiveness, and creating new opportunities for businesses. However, it is crucial to address the challenges and risks associated with technology to ensure inclusive and sustainable trade growth. By harnessing the potential of technology while promoting digital inclusion and cybersecurity, countries can unlock the full benefits of technology in international trade.

Investigating the potential of technology in poverty reduction

Poverty is a complex and multi-dimensional issue that affects millions of people around the world. It is characterized by a lack of sufficient income and resources to meet basic needs, such as food, shelter, education, and healthcare. One of the key factors contributing to poverty is limited access to technology and its benefits. However, technology can also be a powerful tool in addressing poverty and promoting economic development. In this section, we will explore the potential of technology in poverty reduction and examine how it can be harnessed to create positive change.

Understanding the role of technology in poverty reduction

Technology has the potential to directly and indirectly alleviate poverty by improving access to resources and opportunities. When properly utilized, it can enhance productivity, increase efficiency, and create new avenues for economic participation. Here are some ways in which technology can contribute to poverty reduction:

1. Access to Information: Technology, particularly the internet, has revolutionized the way information is accessed and shared. It provides individuals with the ability to acquire knowledge, develop skills, and access educational resources. This can empower individuals to improve their employability and pursue economic opportunities.

2. Enhancing Productivity: Technological advancements have the potential to increase productivity in various sectors. For example, the use of machinery and automation can enhance the efficiency of manufacturing processes, leading to higher output and lower production costs. This can result in increased income and job creation, thus reducing poverty.

3. Digital Financial Inclusion: Technology has played a significant role in improving financial inclusion and access to financial services. Mobile banking, digital payment systems, and microfinance platforms have made it easier for individuals in low-income communities to save, borrow, and manage their finances. This enables them to build assets, manage risks, and invest in income-generating activities.

4. Market Access and Trade: Technology can facilitate market access for small-scale farmers and entrepreneurs, particularly in remote and underserved

areas. E-commerce platforms and digital marketplaces provide opportunities to connect producers with buyers, eliminating intermediaries and reducing transaction costs. This enables small businesses to reach a wider customer base and access competitive markets, thus increasing their income and reducing poverty.

5. Healthcare and Education: Technology can significantly improve access to healthcare services and educational resources in low-income communities. Telemedicine allows individuals in remote areas to receive medical consultations and treatment without the need to travel long distances. Similarly, online educational platforms provide access to quality education and skills training, promoting human capital development and increasing employability.

Barriers to technology adoption in addressing poverty

While technology holds immense potential for poverty reduction, there are several barriers that need to be addressed to ensure its effective adoption and utilization. These barriers include:

1. Infrastructure: Limited access to reliable electricity, internet connectivity, and digital devices hinders the adoption and usage of technology in low-income communities. Lack of infrastructure development in rural and underserved areas prevents individuals from fully benefiting from the opportunities offered by technology.

2. Affordability: The cost of technology and digital devices can be a significant barrier for individuals living in poverty. High initial investment and recurring costs, such as internet subscriptions, may make technology inaccessible to those with limited financial resources.

3. Digital Divide: Unequal access to technology and the internet creates a digital divide between individuals and communities. This divide is often along socio-economic lines, with marginalized groups having limited or no access to technology. This further perpetuates existing inequalities and hinders poverty reduction efforts.

4. Digital Literacy: Lack of digital literacy and skills can limit the effective utilization of technology in poverty reduction. Individuals need to be equipped with the necessary skills to effectively navigate digital platforms, access information, and utilize technology for their economic and social well-being.

Examples of technology-enabled poverty reduction initiatives

Numerous organizations and initiatives are leveraging technology to address poverty and create sustainable development. Here are a few examples:

1. GrameenPhone: In Bangladesh, GrameenPhone, a mobile telecommunications company, has partnered with microfinance institutions to provide access to financial services for individuals in rural areas. Through mobile banking and digital payment systems, individuals can save, borrow, and transfer money, leading to increased financial inclusion and empowerment.

2. M-Kopa Solar: M-Kopa Solar, a Kenyan company, provides affordable solar power solutions to off-grid households. By utilizing mobile-based payment systems, individuals can access clean and affordable energy, improving their quality of life and reducing reliance on costly and polluting alternatives.

3. One Laptop per Child: The One Laptop per Child (OLPC) initiative aims to provide affordable laptops to children in low-income communities worldwide. By equipping children with digital devices and educational resources, the initiative aims to bridge the digital divide and enhance access to quality education.

4. WeFarm: WeFarm is a peer-to-peer knowledge-sharing platform for farmers in developing countries. Through SMS-based technology, farmers can ask questions, seek advice, and share valuable insights with other farmers, improving agricultural practices and increasing productivity.

Conclusion

Technology has the immense potential to contribute to poverty reduction by improving access to resources, enhancing productivity, and creating new opportunities. However, addressing the barriers to technology adoption and ensuring equitable access is crucial to harnessing its full potential. Governments, organizations, and stakeholders need to collaborate to develop and implement policies and programs that promote inclusive and responsible technology adoption. By leveraging technology effectively, we can create a more equitable and prosperous society, where poverty becomes a thing of the past.

Understanding the challenges and risks associated with technology adoption

Technology adoption is the process through which individuals, businesses, and governments embrace and integrate new technological innovations into their operations. While technology adoption offers numerous benefits, it also presents several challenges and risks that need to be carefully considered. In this section, we will explore these challenges and risks associated with technology adoption, and discuss strategies to address them effectively.

Short-term costs and financial barriers

One of the primary challenges of technology adoption is the short-term costs associated with implementation. Upgrading or adopting new technologies often requires significant financial investment, including the purchase of equipment, software licenses, training programs, and hiring technical experts. Small businesses and developing countries may face financial barriers that make it challenging to invest in new technologies. Additionally, organizations may be hesitant to invest in new technologies due to uncertain returns on investment or budget constraints.

To address these challenges, governments and organizations could consider providing financial incentives such as tax breaks, grants, or loans specifically aimed at technology adoption. Collaboration with technology providers to develop affordable and scalable solutions can also help overcome financial barriers. Moreover, fostering an environment of innovation and entrepreneurship can encourage the growth of small and medium-sized enterprises (SMEs) that are more agile in adopting new technologies.

Compatibility and interoperability issues

Another challenge in technology adoption is the compatibility and interoperability issues that arise when integrating new technologies with existing systems. Organizations often rely on legacy systems and infrastructure, which may not be compatible with the new technologies. These compatibility issues can lead to delays, disruptions, and additional costs during the adoption process.

To mitigate these challenges, thorough planning and assessment of existing infrastructure are essential. Organizations should conduct a comprehensive analysis of their current systems to identify potential compatibility issues and develop strategies to address them. Collaboration with technology vendors and engaging in pilot projects can help identify and resolve compatibility issues before full-scale implementation. Additionally, the establishment of industry standards and open-source platforms can improve interoperability between different technologies and facilitate smoother adoption processes.

Resistance to change and workforce retraining

Resistance to change is a common challenge during technology adoption, as it often disrupts established routines and work processes. Employees may resist the adoption of new technologies due to fear of job loss, lack of understanding, or discomfort with learning new skills. In addition, organizations may face resistance from stakeholders who are accustomed to traditional methods and are hesitant to embrace technological advancements.

To overcome resistance to change, organizations should prioritize change management strategies that involve effective communication, employee training, and engagement. This includes providing clear explanations of the benefits of technology adoption, offering opportunities for employee skill development and retraining, and involving employees in the decision-making process. Creating a culture that values innovation and continuous learning can also foster a positive environment for technology adoption.

Data security and privacy concerns

Technology adoption often involves the collection, storage, and analysis of large amounts of data, which raises concerns about data security and privacy. Organizations must implement robust cybersecurity measures to protect sensitive information from unauthorized access, data breaches, and cyber-attacks. Moreover, new technologies such as artificial intelligence and IoT devices introduce additional risks due to their capability to gather and analyze personal data.

To address these concerns, organizations need to prioritize data protection and privacy as part of their technology adoption strategy. This includes implementing encryption, access controls, and regular security audits to safeguard sensitive data. Compliance with relevant data protection regulations and standards should also be prioritized. Furthermore, organizations should educate employees and stakeholders about data privacy best practices to ensure responsible handling of data throughout the adoption process.

Ethical implications and social impact

Technology adoption can have significant ethical implications and social impact. Incorporating new technologies can lead to job displacement, widening the gap between skilled and unskilled workers. Moreover, the use of technologies such as AI and automation raises ethical concerns related to algorithmic bias, privacy infringement, and accountability.

To address these concerns, organizations and policymakers need to take a proactive approach in considering the ethical implications of technology adoption. This involves developing regulations and guidelines that safeguard individuals' rights and promote fairness and inclusivity. Additionally, investing in education and training programs that equip individuals with the necessary skills to adapt to technological advancements is crucial. Collaboration between different stakeholders, including academia, industry leaders, and policymakers, is essential for developing ethical frameworks that ensure responsible and equitable technology adoption.

Case Study: Technology Adoption in the Healthcare Sector

An example of the challenges and risks associated with technology adoption can be observed in the healthcare sector. The integration of electronic health records (EHRs) and telemedicine technologies has the potential to improve healthcare delivery and patient outcomes, but it also presents various challenges.

One challenge is the initial cost of implementing EHR systems and training healthcare professionals to use them effectively. This financial burden can be a barrier for smaller healthcare facilities or those in resource-constrained settings. Furthermore, interoperability issues between different EHR systems can hinder data sharing and coordination of care between different healthcare providers.

To address these challenges, governments and healthcare organizations can offer financial incentives and support to encourage technology adoption. Investing in training programs for healthcare professionals can facilitate the transition to EHRs and ensure their effective use. Additionally, regulatory efforts to establish interoperability standards and data exchange frameworks can improve the sharing of patient information and enhance the overall quality of care.

Conclusion

While technology adoption offers immense potential for economic growth and development, it also comes with its own set of challenges and risks. Organizations and policymakers need to proactively address these challenges to ensure successful technology adoption. By investing in financial incentives, improving compatibility and interoperability, fostering a culture of change and continuous learning, prioritizing data security and privacy, considering the ethical implications, and promoting collaboration between stakeholders, technology adoption can be effectively managed to drive sustainable and equitable development.

Identifying the factors that hinder technology diffusion

Technology diffusion refers to the process by which new technologies are adopted and spread within a society or across different countries. While technology has the potential to bring about significant economic and social benefits, its diffusion is often hindered by various factors. Understanding these factors is crucial for policymakers and stakeholders to develop strategies to overcome them and promote technology adoption on a broader scale. In this section, we will explore some of the key factors that hinder technology diffusion.

1. Lack of Access to Infrastructure

One of the primary factors that hinder technology diffusion is the lack of access to essential infrastructure. Technology requires a robust infrastructure to function effectively. This includes access to reliable electricity, a well-developed telecommunications network, and efficient transportation systems. In many developing countries, inadequate infrastructure poses a significant barrier to technology diffusion. Limited or unreliable access to electricity and the internet, inadequate road networks, and insufficient logistical capabilities can limit the adoption and spread of technology. Without proper infrastructure, it becomes challenging for individuals and businesses to utilize and benefit from technological advancements.

2. High Initial Costs

Another factor that impedes technology diffusion is the high initial costs associated with adopting new technologies. Many innovative technologies require significant upfront investments in equipment, training, and infrastructure. These costs can be particularly burdensome for small and medium-sized enterprises (SMEs) and individuals in low-income communities. Moreover, economies of scale often play a role in reducing the costs of technology over time. As a result, early adopters may face higher costs compared to those who adopt the technology at a later stage. The high initial costs associated with technology adoption can deter potential adopters, especially in resource-constrained settings.

3. Limited Technological Literacy and Skills

The lack of technological literacy and skills is another critical factor hindering technology diffusion. Effective utilization of technology requires individuals to have the necessary knowledge and skills. However, many individuals, especially in developing countries, lack the technical skills required to operate and benefit from new technologies. This lack of technological literacy can be attributed to factors such as limited access to quality education, inadequate training opportunities, and a lack of awareness about the potential benefits of technology. Without the necessary skills, individuals and organizations may struggle to adopt and integrate technology into their daily lives and operations.

4. Regulatory Barriers and Policies

Regulatory barriers and policies can significantly hinder technology diffusion. Overly restrictive regulations, bureaucratic processes, and legal complexities can create obstacles to the adoption of new technologies. In some cases, outdated or rigid regulations may not be able to keep pace with technological advancements, making it challenging for innovators to bring new technologies to market. Moreover, the lack of coherent policies and frameworks for technology adoption can create uncertainty and discourage investment in innovative technologies. To facilitate technology diffusion, it is essential for governments to establish an enabling regulatory environment that promotes innovation, protects intellectual property rights, and removes unnecessary barriers to adoption.

5. Cultural and Social Factors

Cultural and social factors also play a significant role in hindering technology diffusion. Attitudes, beliefs, and social norms can influence the adoption of new technologies. Skepticism, resistance to change, and fear of job displacement can deter individuals and organizations from embracing technology. Cultural factors such as a preference for traditional practices or a lack of awareness about the potential benefits of technology can further impede diffusion. Addressing these cultural and social barriers requires targeted awareness and education campaigns to cultivate a positive mindset towards technology adoption and highlight its potential benefits.

6. Lack of Access to Financing

Access to financing is a critical factor that hinders technology diffusion, particularly for resource-constrained individuals and businesses. Many innovative technologies require substantial financial resources for research and development, production, and marketing. However, access to financing options such as loans, venture capital, or grants may be limited, particularly in developing countries. The lack of financial resources and supportive mechanisms can prevent innovators and entrepreneurs from commercializing and scaling their technological innovations. To overcome this barrier, governments and other stakeholders need to create mechanisms that provide affordable financing options and support innovative startups and SMEs.

7. Knowledge and Information Gaps

Knowledge and information gaps hinder technology diffusion by limiting access to critical information about new technologies. Lack of awareness about the availability and potential benefits of innovative technologies can hamper adoption rates. Dissemination of accurate and up-to-date information about new technologies, their functionalities, and case studies demonstrating their successful implementation is crucial for promoting technology diffusion. Bridging the knowledge and information gaps through targeted educational programs, training initiatives, and awareness campaigns can help overcome this barrier.

8. Geographic and Spatial Factors

In some cases, technology diffusion can be hindered by geographic and spatial factors. Remote and rural areas often face challenges in accessing technology due to a lack of physical infrastructure, limited connectivity, and difficulties in reaching marginalized communities. The digital divide, which refers to the gap in internet and digital access between different regions and socio-economic groups, exacerbates these challenges. Bridging the digital divide through initiatives such as expanding internet connectivity and providing technology resources to underserved areas can help overcome these geographic and spatial barriers.

9. Political and Institutional Factors

Political and institutional factors also play a crucial role in hindering technology diffusion. Inadequate governance structures, corruption, and political instability can create an unfavorable environment for technology adoption. Lack of political commitment to promoting technology diffusion, weak institutional support, and inadequate enforcement of intellectual property rights can deter innovators and investors. Creating a stable political and institutional environment that fosters innovation, protects intellectual property, and provides support for technology adoption is essential for overcoming these barriers.

10. Environmental and Sustainability Considerations

Environmental and sustainability considerations can also hinder technology diffusion. Some innovative technologies may face resistance due to concerns about their environmental impact. Uncertainty about the long-term sustainability and potential risks associated with new technologies can create skepticism and resistance among individuals and organizations. Addressing these concerns requires robust environmental assessment frameworks, transparency in technology development and deployment, and proactive communication about the measures taken to ensure sustainability and minimize adverse impacts.

In summary, identifying and understanding the factors that hinder technology diffusion is crucial for policymakers and stakeholders aiming to promote technology adoption on a broader scale. The lack of access to infrastructure, high initial costs, limited technological literacy, regulatory barriers, cultural and social factors, lack of access to financing, knowledge and information gaps, geographic and spatial factors, political and institutional factors, and environmental considerations all contribute to hindrances in technology diffusion. Overcoming these barriers requires the formulation of targeted policies, investments in education and training, improving access to financing, addressing cultural and social barriers, promoting sustainable technology solutions, and creating an enabling environment for technology adoption and diffusion.

Discussing the Ethical Implications of Technological Advancements

The rapid advancements in technology bring numerous benefits to society, such as improved communication, increased productivity, and enhanced quality of life. However, alongside these benefits, there are also ethical concerns that arise from the development and deployment of new technologies. In this section, we will explore some of the key ethical implications of technological advancements and discuss the challenges they present.

Privacy and Data Protection

One major ethical concern related to technological advancements is the issue of privacy and data protection. With the increasing use of digital technologies, individuals are generating and sharing vast amounts of personal data. This data can be collected, analyzed, and used by various entities, including governments and corporations. The ethical question arises: how can we ensure the protection of individuals' privacy in an era of pervasive data collection and surveillance?

To address this concern, it is imperative to develop robust legal and regulatory frameworks to safeguard individuals' privacy rights. These frameworks should include measures that promote transparency and informed consent in data collection practices. Additionally, organizations should adopt strong security measures to protect personal data from unauthorized access or breach. It is also crucial to educate individuals about their rights and provide them with tools to control how their data is collected and used.

Algorithmic Bias and Fairness

Another ethical issue stemming from technological advancements is algorithmic bias and fairness. As artificial intelligence (AI) systems become increasingly prevalent, they are being used to make decisions that impact individuals' lives, such as hiring, credit scoring, and criminal justice. However, these AI systems can inherit biases from the data they are trained on, leading to discriminatory outcomes.

To address algorithmic bias, it is essential to ensure that AI models are trained on diverse and representative datasets. This includes considering factors such as race, gender, and socioeconomic background to mitigate biases that may arise. Additionally, there should be transparency in the decision-making process of AI systems, enabling individuals to understand how and why certain decisions are made. Ongoing monitoring and evaluation of AI systems can help identify and correct biases as they emerge.

Job Displacement and Economic Inequality

Technological advancements, particularly in automation and artificial intelligence, have the potential to disrupt labor markets and lead to job displacement. This raises ethical concerns regarding the impact on individuals whose jobs are replaced by machines. Moreover, it can exacerbate existing economic inequalities, as those who possess the skills to work alongside or operate these technologies may benefit while others are left unemployed.

To address the ethical implications of job displacement, it is crucial to foster education and retraining programs that equip individuals with the skills needed for the jobs of the future. Governments and organizations should invest in lifelong learning initiatives to ensure individuals can adapt to changing technological landscapes. Moreover, social safety nets and policies that promote income redistribution may be necessary to mitigate the negative impacts of economic inequality.

Autonomous Systems and Accountability

Advancements in autonomous systems, such as self-driving cars and unmanned aerial vehicles, introduce ethical questions regarding accountability. When accidents occur involving these systems, who should be held responsible? Should it be the developers, manufacturers, or the AI systems themselves? The challenge lies in defining liability and ensuring appropriate mechanisms for accountability in the face of complex and autonomous technologies.

To address this issue, legal and regulatory frameworks should be established to assign responsibility and liability for accidents involving autonomous systems. These frameworks should consider aspects such as the level of autonomy, the role of human oversight, and the quality of the underlying technology. Additionally, it may be necessary to develop safety standards and certification processes to ensure the reliability and accountability of autonomous systems.

Environmental Impact

Technological advancements can also have significant environmental implications. For example, the manufacturing and disposal of electronic devices contribute to electronic waste, while the energy consumption of data centers and computing infrastructure contributes to carbon emissions. Ethical considerations arise concerning sustainable and responsible use of technology to mitigate environmental harm.

To address the environmental impact of technological advancements, it is essential to promote eco-friendly design and manufacturing practices. This includes adopting sustainable materials, improving energy efficiency, and facilitating recycling and proper disposal of electronic waste. Furthermore, there should be an emphasis on developing and promoting environmentally friendly technologies, such as renewable energy sources and energy-efficient systems.

In conclusion, technological advancements bring about numerous ethical implications that need to be carefully considered. Privacy and data protection, algorithmic bias, job displacement, accountability of autonomous systems, and environmental impact are just a few of the ethical concerns that arise in the context of technological advancements. It is imperative to develop and implement ethical frameworks, policies, and practices that ensure the responsible and beneficial use of technology for the betterment of society. By addressing these ethical challenges, we can ensure a more equitable, inclusive, and sustainable future driven by technological advancements.

Key Terms and Concepts

- Privacy and data protection

- Algorithmic bias and fairness

- Job displacement and economic inequality

- Autonomous systems and accountability
- Environmental impact

Further Reading

1. Floridi, L. (2016). "The Ethics of Information." Oxford University Press.
2. Bostrom, N. and Roache, R. (2016). "Ethical Guidelines for Autonomous Vehicles." Nature.
3. Taddeo, M. and Floridi, L. (2018). "Ethics of AI and Robotics." Stanford Encyclopedia of Philosophy.

Discussion Questions

1. How can organizations balance the need for data collection and analysis with individuals' right to privacy?
2. What steps can be taken to mitigate algorithmic bias and ensure fairness in AI systems?
3. How should society address the ethical implications of job displacement caused by technological advancements?
4. Which entities should be held accountable for accidents involving autonomous systems?
5. What strategies can be implemented to minimize the environmental impact of technological advancements?

Remember to critically reflect on these topics and engage in respectful discussions with your classmates or colleagues.

Historical Overview of Technological Transformations

Examining the Industrial Revolution and its impact on economic development

The Industrial Revolution was a period of significant technological advancements, which occurred in the late 18th to early 19th century. It marked a shift from manual labor to machine-based manufacturing processes, transforming various industries such as textiles, iron and coal mining, transportation, and agriculture. This section will explore the key elements of the Industrial Revolution and its profound impact on economic development.

Understanding the Industrial Revolution

The Industrial Revolution originated in Great Britain and gradually spread to other parts of Europe, North America, and eventually the rest of the world. It was characterized by several key developments:

- **Mechanization:** The introduction of new machinery and the replacement of manual labor with mechanical processes. For example, the invention of the steam engine by James Watt revolutionized transportation and factory production.

- **Mass production:** The ability to produce goods on a large scale, thanks to the utilization of machinery and the division of labor. This led to increased productivity and the availability of affordable goods.

- **Urbanization:** The movement of people from rural areas to cities in search of employment opportunities in factories. This rapid urbanization resulted in the growth of industrial towns and the emergence of new social and economic dynamics.

- **Technological innovations:** The Industrial Revolution was fueled by a series of breakthrough inventions, including the spinning jenny, power loom, cotton gin, and the Bessemer process for steel production, among others. These innovations significantly transformed various industries and accelerated economic growth.

Impact on Economic Development

The Industrial Revolution had a profound and far-reaching impact on economic development. It brought about significant changes in the following aspects:

1. **Increased productivity:** The introduction of machinery and mass production techniques drastically increased productivity in industries. This led to greater output and efficiency, contributing to economic growth.

2. **Expanding industries:** The development of new technologies and the growth of industries such as textiles, iron and steel, and transportation

stimulated economic expansion. This expansion created job opportunities, attracted investments, and spurred further innovation.

3. **Improved living standards:** The Industrial Revolution resulted in higher wages, increased employment opportunities, and the availability of affordable goods. This improvement in living standards had a positive impact on the overall well-being of the population.

4. **Urbanization and social changes:** The shift from agrarian societies to urban industrial centers led to significant societal changes. Urban areas became hubs of economic activity, cultural exchange, and social mobility, as people flocked to cities for employment. However, urbanization also brought challenges, such as overcrowding, poor working conditions, and social inequalities.

5. **Global trade and colonialism:** The Industrial Revolution enabled the expansion of global trade networks, as countries sought new markets for their manufactured goods and sources of raw materials. This drive for resources fueled colonialism and gave rise to economic dominance by certain nations.

6. **Technological advancements:** The inventions and technological advancements of the Industrial Revolution laid the foundation for further innovations in subsequent centuries. For example, the steam engine paved the way for the development of railways and the growth of the transportation sector.

Challenges and Criticisms

While the Industrial Revolution brought significant economic benefits, it also had its share of challenges and criticisms. Some of the main concerns include:

- **Working conditions and labor exploitation:** The early years of industrialization saw harsh working conditions, long hours, and low wages for workers, especially in factories and mines. This led to social unrest and the emergence of labor movements advocating for workers' rights.

- **Environmental impact:** The Industrial Revolution marked a turning point in human impact on the environment. Rapid industrialization led to pollution, deforestation, and resource depletion. The long-term consequences of these environmental changes are still felt today.

- **Social inequalities:** The unequal distribution of wealth and the emergence of industrial capitalists and a working class created social divisions. This income disparity and the concentration of wealth in the hands of a few sparked debates on social justice and income equality.

- **Displacement of traditional industries:** The transition from traditional craftsmanship to factory-based production displaced many skilled workers and disrupted traditional industries in rural areas. This led to economic upheaval and social dislocation in certain regions.

Industrial Revolution in Context

It is essential to understand the Industrial Revolution in the broader context of historical and economic development. The advancements and transformations brought about during this period laid the groundwork for subsequent technological revolutions and economic growth. The lessons learned from the early industrialization experience have shaped policies and approaches to mitigate the negative impacts and maximize the benefits of subsequent technological evolutions.

Conclusion

The Industrial Revolution was a pivotal period in human history that brought about significant technological, economic, and social changes. It marked a shift towards industrialization, mechanization, and mass production, transforming not only industries but also society as a whole. While the Industrial Revolution contributed to economic growth and improved living standards, it also brought about various challenges and criticisms. Understanding this historical context is crucial for grasping the subsequent technological advancements and their impact on economic development. As we delve into further sections of this book, we will explore how subsequent technological revolutions have built upon the foundations laid during the Industrial Revolution.

Analyzing the role of electricity in shaping modern economies

Electricity is a fundamental form of energy that plays a crucial role in shaping modern economies. Its widespread availability and use have transformed various sectors, including industry, transportation, communication, and household activities. In this section, we will explore the impact of electricity on economic development, analyze its role in different sectors, and discuss its implications for sustainable growth.

The Impact of Electricity on Economic Development

The availability of reliable and affordable electricity is a critical factor in economic development. It enables the growth and expansion of industries by providing a consistent power supply for manufacturing processes. With electricity, machines and equipment can be operated efficiently, leading to increased productivity and output. This, in turn, drives economic growth and creates employment opportunities.

Moreover, electricity facilitates the development of new industries and technologies. For example, the invention of the electric motor revolutionized manufacturing processes, leading to the mass production of goods and the rise of the assembly line. Electricity also powered the development of various technological innovations, such as electric lighting, refrigeration, and telecommunication systems.

In addition to industrial development, electricity has a profound impact on the service sector. It fuels advancements in communication technology, enabling faster and more efficient transmission of information, which is vital for modern businesses and organizations. Furthermore, electricity drives the growth of the transportation sector, allowing the widespread use of electric vehicles and the development of electric-powered public transportation systems.

Overall, the availability of electricity is a key driver of economic development. It fuels innovation, enhances productivity, and enables the efficient operation of various industries and sectors.

Electricity in Different Sectors

1. Industrial Sector: The industrial sector is one of the primary beneficiaries of electricity. It powers machinery, equipment, and tools used in manufacturing and processing activities. Electricity enables automation, leading to increased production efficiency and cost-effectiveness. It also provides a reliable source of power for industries, reducing dependence on fossil fuels and promoting environmental sustainability.

2. Transportation Sector: The adoption of electric vehicles (EVs) has gained significant momentum in recent years, driven by advancements in battery technology and environmental concerns. Electric-powered cars, buses, and trains offer a cleaner and more sustainable alternative to traditional combustion engines, reducing greenhouse gas emissions and dependence on fossil fuels. Furthermore, electric transportation systems, such as electric trains and trams, provide efficient and eco-friendly modes of transport in urban areas.

3. Communication Sector: The communication sector heavily relies on electricity to power a wide range of devices and infrastructure. Telecommunication networks, data centers, and mobile devices all require electricity to function. Additionally, electricity enables the transmission and distribution of communication signals, allowing for seamless connectivity and information exchange globally.

4. Household Sector: Electricity plays a vital role in households, powering essential appliances and devices. It provides lighting, heating, and cooling systems, as well as the energy required for cooking, refrigeration, and entertainment. Access to electricity in households improves living standards, enhances productivity, and enables the use of modern technologies that simplify daily tasks and improve the quality of life.

Implications for Sustainable Growth

While electricity has brought numerous benefits to modern economies, its production and consumption have implications for sustainable growth. The generation of electricity from fossil fuels contributes to air pollution, climate change, and resource depletion. Therefore, transitioning to cleaner and renewable sources of energy is crucial for long-term sustainability.

Renewable energy sources, such as solar, wind, hydro, and geothermal power, offer environmentally friendly alternatives to conventional electricity generation. By investing in renewable energy infrastructure, countries can reduce carbon emissions and mitigate the impacts of climate change. Additionally, decentralized renewable energy systems enable energy access in remote and underserved areas, promoting inclusive and equitable development.

Moreover, the adoption of energy-efficient technologies and practices is essential in minimizing energy waste and optimizing electricity consumption. Energy-efficient appliances, building insulation, and smart grid systems are examples of measures that can significantly reduce electricity consumption while maintaining economic productivity.

Promoting sustainable electricity use also requires raising awareness and educating individuals about the environmental and social impacts of their energy consumption habits. Government policies and incentives, such as carbon pricing and subsidies for renewable energy projects, can accelerate the transition to sustainable electricity generation and consumption.

In conclusion, electricity is a vital component in shaping modern economies. Its availability and use have brought about significant advancements in various sectors, including industry, transportation, communication, and households.

However, ensuring sustainable growth requires transitioning to cleaner and renewable sources of energy and adopting energy-efficient practices. By doing so, we can harness the full potential of electricity while minimizing its environmental impact and promoting sustainable economic development.

Discussing the advent of the computer and its impact on businesses

The advent of the computer has had a profound impact on businesses across various industries. In this section, we will explore the historical background of the computer, its key features, and the ways in which it has revolutionized the business landscape.

Historical Background

The development of computers can be traced back to the mid-20th century, with significant contributions from pioneers such as Alan Turing, John von Neumann, and Grace Hopper. Early computers were large, room-sized machines that were primarily used by government agencies and research institutions for complex calculations.

One notable milestone in the history of computers was the invention of the integrated circuit in the late 1950s. This breakthrough allowed for the miniaturization of computer components, making computers smaller, faster, and more affordable. By the 1970s, personal computers (PCs) started to enter the market, paving the way for widespread computer adoption in businesses and households.

Key Features of Computers

Computers are programmable devices that can store, retrieve, and process information. They are characterized by the following key features:

- **Processing Power:** Computers have the ability to perform complex calculations and execute instructions at high speeds. This enables businesses to automate repetitive tasks, analyze large datasets, and run sophisticated simulations, among other applications.

- **Storage Capacity:** Computers can store vast amounts of data, ranging from text documents and images to videos and databases. This allows businesses to organize and access information efficiently, leading to improved decision-making processes.

- **Connectivity:** Through the internet and other networks, computers can establish connections with other computers and devices worldwide. This enables businesses to communicate, collaborate, and share information in real-time, transcending geographical boundaries.

- **Software Capabilities:** Computers operate using software programs that enable users to perform various tasks. From word processors and spreadsheets to specialized software for accounting, project management, and customer relationship management, computers offer a plethora of applications tailored to meet business needs.

Impact on Businesses

The advent of computers has had a transformative impact on businesses, revolutionizing various aspects of their operations. Let's explore some key areas where computers have made a significant difference:

Automation and Efficiency Computers have enabled businesses to automate repetitive and time-consuming tasks, leading to increased efficiency and productivity. For example, with the help of computer-controlled manufacturing systems, factories can produce goods at a faster rate and with higher precision. Similarly, office automation tools, such as email, word processing software, and spreadsheets, have streamlined administrative tasks, allowing employees to focus on more strategic activities.

Data Processing and Analysis The ability of computers to process and analyze large volumes of data has revolutionized decision-making processes in businesses. With the help of computer algorithms and data analytics tools, organizations can extract valuable insights from datasets, enabling them to make data-driven decisions. This has led to improvements in areas such as marketing campaigns, supply chain management, and customer segmentation.

Communication and Collaboration Computers have transformed communication and collaboration within businesses. With the advent of email, instant messaging, and video conferencing tools, employees can connect with colleagues, partners, and clients around the globe in real-time. This has accelerated decision-making processes, enhanced teamwork, and facilitated global business operations.

Access to Information Computers have democratized access to information, empowering businesses with vast knowledge resources. Through the internet, businesses can access market research, industry trends, competitor analysis, and other valuable information that can inform strategic decision-making. This has leveled the playing field, enabling small businesses to compete with larger corporations on a global scale.

E-commerce and Online Business The rise of computers and the internet has given birth to e-commerce, enabling businesses to sell products and services online. This has opened up new markets and revenue streams, allowing businesses to reach customers globally. Additionally, computers have facilitated the growth of online business models, such as software-as-a-service (SaaS) and platform-based businesses, further expanding opportunities for entrepreneurship.

Challenges and Adaptation

While the advent of computers has brought numerous benefits to businesses, it has also posed challenges that require adaptation. Some of these challenges include:

- **Cybersecurity:** As businesses rely more and more on computer systems, the risk of cyber threats, such as data breaches and malicious attacks, increases. Businesses need to invest in robust cybersecurity measures to protect their sensitive information and digital assets.

- **Skill Development:** With the rapid pace of technological advancements, businesses need skilled employees who can effectively leverage computer technologies. Upskilling and reskilling programs are crucial to ensure that employees have the necessary digital literacy and technical expertise to adapt to changing work environments.

- **Digital Divide:** The digital divide refers to the gap between those who have access to computers and the internet and those who do not. Businesses need to address this divide by promoting digital inclusion initiatives, ensuring equal access to opportunities for all segments of society.

- **Privacy and Ethical Concerns:** The extensive use of computers is accompanied by privacy and ethical considerations. Businesses must navigate privacy regulations, protect consumer data, and address ethical concerns related to the use of technologies such as artificial intelligence and big data analytics.

Example: Impact on Retail Industry

To illustrate the impact of computers on businesses, let's consider the retail industry. Computers and technology have transformed various aspects of retail operations, from inventory management to customer experience.

One key application of computers in retail is the use of point-of-sale (POS) systems. These systems allow retailers to automate sales transactions, manage inventory in real-time, and generate valuable sales reports. By replacing manual cash registers, computers have improved checkout processes, reduced human errors, and facilitated efficient inventory management.

Moreover, computers have enabled retailers to enhance the customer experience through online shopping platforms and personalized marketing campaigns. E-commerce websites, powered by computers, have made it easier for customers to browse and purchase products online, leading to increased convenience and access to a wider range of products. Additionally, computers enable retailers to analyze customer data and preferences, allowing them to tailor marketing strategies and provide personalized recommendations.

In summary, the advent of computers has significantly impacted businesses, including the retail industry. By automating tasks, improving data processing capabilities, enhancing communication, and enabling online business models, computers have revolutionized the way businesses operate. However, businesses must also address challenges related to cybersecurity, skill development, digital divide, and privacy concerns to fully leverage the potential of computers for growth and success.

Examining the rise of the internet and the digital revolution

The rise of the internet and the subsequent digital revolution have had a profound impact on the global economy. This section will explore the evolution of the internet, the transformative power of digital technologies, and the implications for economic development.

1. Background The concept of the internet originated in the late 1960s as a decentralized network called ARPANET, developed by the United States Department of Defense. Initially, its purpose was to facilitate communication and data transfer between research institutions and universities. However, with advancements in technology and the introduction of the World Wide Web in the 1990s, the internet became accessible for commercial and public use, leading to a revolution in information exchange and connectivity.

2. The Digital Revolution The digital revolution refers to the extensive use of digital technologies, such as computers, smartphones, and the internet, to transform various aspects of society. This section focuses on the economic implications of this revolution.

2.1 Connectivity and Access The internet has revolutionized global connectivity, enabling people around the world to connect with each other and access vast amounts of information. The widespread availability of affordable internet access has significantly reduced the barriers to entry for individuals and businesses, making it easier to participate in the global digital economy.

2.2 E-commerce and Online Marketplaces The rise of the internet has facilitated the growth of e-commerce, allowing businesses to sell products and services online. Online marketplaces, such as Amazon and Alibaba, have disrupted traditional retail models, providing consumers with a wider range of products and increased convenience. E-commerce has also enabled small and medium-sized enterprises to access global markets, democratizing international trade.

2.3 Digital Platforms and Sharing Economy Digital platforms, such as Uber and Airbnb, have transformed various industries through the sharing economy model. These platforms connect individuals who want to offer or consume goods and services, creating new opportunities for entrepreneurship and income generation. The sharing economy has also led to concerns regarding labor rights and regulatory challenges.

2.4 Data and Analytics The digital revolution has led to the proliferation of data generated from various sources, including social media, sensors, and online transactions. Advanced analytics and machine learning techniques enable businesses and governments to derive valuable insights from these data streams. This has revolutionized decision-making processes and improved efficiency in sectors such as marketing, healthcare, and urban planning.

3. Implications for Economic Development The rise of the internet and the digital revolution have presented both opportunities and challenges for economic development.

3.1 Increased Productivity and Efficiency Digital technologies have the potential to significantly enhance productivity and efficiency across sectors. Automation and digitization of processes can streamline operations, reduce costs, and improve the quality of goods and services. For example, cloud computing allows businesses to access computing resources on-demand, eliminating the need for extensive IT infrastructure.

3.2 Job Displacement and Creation While digital technologies have created new job opportunities, they have also resulted in job displacement, particularly for routine and repetitive tasks. The automation of certain jobs may lead to short-term disruptions in the labor market. However, historically, technological advancements have led to the creation of new jobs that require higher skills and often lead to overall increased employment in the long run.

3.3 Digital Divide and Inequality The digital revolution has exacerbated existing inequalities, creating a digital divide between those who have access to digital technologies and those who don't. This divide can perpetuate socioeconomic disparities, as people without access to digital technologies are at a disadvantage in terms of education, employment, and economic opportunities. Bridging this divide requires targeted policies and initiatives to ensure equal access and digital literacy for all.

3.4 Innovation and Entrepreneurship The digital revolution has democratized innovation, enabling individuals and small businesses to develop and scale innovative products and services with minimal upfront costs. Startups can leverage digital platforms and global connectivity to reach customers worldwide, reducing traditional barriers to entry. However, the rapid pace of technological change can also create challenges for startups, who must continuously adapt to remain competitive.

4. Policy and Regulation The internet and the digital revolution have raised complex policy and regulatory challenges. Governments around the world are grappling with issues related to privacy, cybersecurity, intellectual property, and fair competition in the digital realm. Balancing innovation and consumer protection is a delicate task that requires proactive policies and international collaborations.

5. Conclusion The rise of the internet and the digital revolution have transformed economies and societies worldwide. The widespread adoption of digital technologies has brought unprecedented connectivity, access to information, and opportunities for economic development. However, it has also presented challenges related to inequality, job displacement, and regulatory issues. Harnessing the potential of the digital revolution requires a comprehensive approach that balances innovation with social and economic considerations.

Investigating recent technological advancements and their implications on the economy

In recent years, technological advancements have been transforming various sectors of the economy, leading to significant changes in productivity, innovation, and job creation. This section explores some of the recent technological advancements and their implications on the economy. We will examine the impact of these advancements on different industries and discuss the opportunities and challenges they present.

Advancements in Artificial Intelligence

Artificial Intelligence (AI) has emerged as a transformative technology, with wide-ranging applications across industries. AI technologies, such as machine learning and natural language processing, have the potential to automate tasks, optimize processes, and improve decision-making.

One of the main implications of AI advancements is increased productivity. AI-powered systems can perform complex tasks that would otherwise require significant time and resources. For example, in the healthcare industry, AI algorithms can analyze medical images and detect abnormalities more accurately and quickly than human experts. This reduces diagnosis time and enhances patient care.

However, the adoption of AI also raises concerns about job displacement. As AI systems become more sophisticated, there is a risk of certain job sectors being automated, leading to unemployment for those who were previously engaged in such tasks. It is crucial that policymakers and organizations focus on reskilling and upskilling the workforce to prepare them for the changing job landscape.

Emerging Technologies in Renewable Energy

The development of renewable energy technologies, such as solar and wind power, has gained significant momentum in recent years. These technologies have several implications for the economy, particularly in the energy sector.

Renewable energy technologies play a crucial role in reducing carbon emissions and mitigating climate change. By shifting to clean energy sources, countries can decrease their reliance on fossil fuels and improve air quality. This transition also presents economic opportunities in terms of job creation and innovation. For instance, the installation and maintenance of solar panels create employment opportunities in the renewable energy industry.

However, the widespread adoption of renewable energy technologies also comes with challenges. The intermittent nature of solar and wind power generation requires effective energy storage solutions to ensure a stable supply. Additionally, the initial costs of implementing renewable energy systems can be high, necessitating government support and policy incentives.

Digital Transformation in Industries

The digital revolution has brought about a rapid transformation in various industries, including manufacturing, finance, and retail. The integration of digital technologies and data analytics has enabled organizations to enhance their efficiency, customer experience, and decision-making.

One key implication of the digital transformation is improved productivity. With the automation of processes and the use of data analytics, organizations can streamline operations, reduce costs, and increase output. For example, in the manufacturing sector, the adoption of advanced manufacturing technologies, such as 3D printing and robotics, has led to faster production cycles and higher product quality.

However, the digital transformation also presents challenges related to cybersecurity and privacy. As organizations become more reliant on digital technologies, the risk of cyber-attacks and data breaches increases. Therefore, it is crucial for organizations to invest in robust cybersecurity measures and comply with privacy regulations.

Implications for the Labor Market

The advancements in technology have significant implications for the labor market. While technology adoption can result in job displacement in some sectors, it also creates new job opportunities in others.

For example, the rise of e-commerce has led to the creation of jobs in logistics and delivery services. Additionally, the demand for technology professionals, such as data scientists and cybersecurity experts, has surged with the increasing adoption of AI and digital technologies.

However, there is a growing concern about the polarization of the labor market due to technological advancements. Skilled workers who can adapt to the changing technological landscape are likely to benefit from new job opportunities and higher wages. Meanwhile, low-skilled workers may face challenges in finding employment as routine tasks become automated.

To address the implications for the labor market, policymakers should focus on promoting lifelong learning and reskilling programs. This will enable workers to adapt to technology-driven changes and acquire the skills needed for the jobs of the future.

In conclusion, recent technological advancements have had a profound impact on the economy. AI technologies have the potential to enhance productivity but also raise concerns about job displacement. Renewable energy technologies present opportunities for clean energy and job creation, while the digital transformation improves efficiency and customer experience. However, challenges related to cybersecurity and labor market polarization need to be addressed. Policymakers, organizations, and individuals need to navigate these advancements carefully to ensure inclusive and sustainable economic growth.

Exploring the potential of emerging technologies in shaping future economies

Emerging technologies have the potential to revolutionize various sectors of the economy, driving innovation, productivity, and economic growth. In this section, we will explore some of the key emerging technologies and their implications for shaping future economies.

Artificial Intelligence and Machine Learning

Artificial Intelligence (AI) and Machine Learning (ML) technologies have already started to transform various industries, and their potential for future economic growth is vast. AI refers to the development of computer systems capable of performing tasks that typically require human intelligence, such as speech recognition, decision-making, and problem-solving. ML, a subset of AI, involves the development of algorithms that allow computer systems to learn and improve from data without explicit programming.

The impact of AI and ML on future economies is significant. These technologies have the potential to automate routine and repetitive tasks, freeing up human resources to focus on more complex and value-added activities. For example, in manufacturing, AI-powered robots can improve efficiency, reduce costs, and enhance product quality. In healthcare, AI algorithms can help diagnose diseases, personalize treatments, and improve patient outcomes. In transportation, self-driving cars and drones powered by AI can revolutionize logistics and mobility.

However, the widespread adoption of AI and ML also raises ethical concerns, such as privacy, bias, and job displacement. It is crucial to develop frameworks and policies that ensure responsible and inclusive deployment of these technologies, addressing societal, ethical, and legal implications.

Internet of Things

The Internet of Things (IoT) refers to the network of interconnected physical devices, vehicles, and appliances embedded with sensors and software that enable them to collect and exchange data over the internet. The IoT has the potential to transform industries and create new economic opportunities.

In future economies, the IoT can enable seamless connectivity and automation across various sectors. For example, in agriculture, IoT devices can monitor soil conditions, track livestock, and optimize irrigation, leading to increased productivity and sustainability. In cities, IoT-enabled smart grids, sensors, and data analytics can improve energy efficiency, enhance public safety, and optimize traffic flow. In healthcare, IoT-enabled devices can enable remote patient monitoring, personalized medicine, and early disease detection.

However, the widespread deployment of IoT also raises concerns regarding data security and privacy. As more devices become interconnected, the need for robust cybersecurity measures to protect sensitive data becomes crucial. Additionally, policies and regulations addressing data ownership, consent, and usage must be put in place to ensure the responsible and ethical implementation of IoT technologies.

Blockchain Technology

Blockchain technology, originally introduced as the underlying technology for cryptocurrencies like Bitcoin, has the potential to transform various industries by providing secure, transparent, and decentralized systems for recording and verifying transactions.

In future economies, blockchain technology can revolutionize sectors such as finance, supply chain management, and healthcare. The decentralized nature of blockchain ensures transparency, reduces the risk of fraud, and enhances trust. For example, in finance, blockchain-based digital currencies and smart contracts can streamline cross-border transactions, reduce costs, and provide financial inclusion to the unbanked population. In supply chain management, blockchain can enable end-to-end traceability, ensuring product authenticity, and reducing counterfeiting. In healthcare, blockchain can securely store and share patient records, facilitate interoperability, and improve data security.

However, the adoption of blockchain technology also faces challenges, such as scalability, interoperability, and regulatory frameworks. Overcoming these challenges and developing standards for interoperability and data privacy will be crucial for the widespread adoption of blockchain technology.

HISTORICAL OVERVIEW OF TECHNOLOGICAL TRANSFORMATIONS

Renewable Energy Technologies

As the world shifts toward cleaner and more sustainable energy sources, renewable energy technologies are expected to play a vital role in shaping future economies. Technologies such as solar, wind, tidal, and geothermal power offer alternatives to traditional fossil fuel-based energy generation.

The potential of renewable energy technologies lies in their ability to reduce carbon emissions, mitigate climate change, and provide energy security. In future economies, renewable energy sources have the potential to create new jobs, drive innovation, and spur economic growth. For instance, the growth of the solar industry has already led to significant job creation, especially in installation and manufacturing.

However, the integration of renewable energy sources into existing energy infrastructure presents challenges, including intermittency, storage, and grid management. Overcoming these challenges requires investment in research and development, innovation in storage technologies, and the development of smart grid systems.

In conclusion, emerging technologies such as AI and ML, IoT, blockchain, and renewable energy technologies have immense potential to shape future economies. While these technologies offer numerous benefits, careful consideration of ethical, regulatory, and policy frameworks is essential to ensure their responsible and inclusive deployment. By embracing and harnessing the potential of emerging technologies, future economies can thrive, driving sustainable development and innovation.

Analyzing the role of artificial intelligence and automation in economic growth

Artificial intelligence (AI) and automation have emerged as major drivers of economic growth in recent years. The application of AI technologies and automation has significantly transformed various sectors, leading to increased productivity, efficiency, and innovation. In this section, we will analyze the role of AI and automation in economic growth, discussing the implications, opportunities, and challenges associated with their adoption.

Understanding Artificial Intelligence

Artificial intelligence refers to the development of computer systems that can perform tasks that typically require human intelligence, such as visual perception, speech recognition, decision-making, and problem-solving. AI encompasses various techniques, including machine learning, neural networks, natural language processing, and robotics. These technologies enable machines to learn, adapt, and improve their performance based on data and experience.

Implications of AI and Automation

The widespread adoption of AI and automation has significant implications for economic growth. These technologies can enhance productivity by automating repetitive and mundane tasks, allowing employees to focus on more complex and strategic activities. Increased productivity can lead to higher output levels and economic expansion.

Moreover, AI and automation can drive innovation by enabling the development of new products, services, and business models. By analyzing large datasets, AI systems can identify patterns, trends, and insights that humans might miss, leading to more effective decision-making and the creation of novel solutions.

Furthermore, AI and automation have the potential to improve the quality and efficiency of processes across various sectors. For example, in manufacturing, robots can perform tasks with precision and speed, resulting in higher product quality and reduced production time.

Opportunities of AI and Automation

The adoption of AI and automation offers several opportunities for economic growth:

1. Increased Efficiency: AI technologies can optimize processes and workflows, leading to improved resource allocation, reduced costs, and enhanced operational efficiency.

2. Enhanced Productivity: Automation of routine tasks frees up human resources, enabling them to focus on higher-value activities, such as innovation, creativity, and problem-solving.

3. Improved Decision-Making: AI systems can analyze vast amounts of data, providing businesses with actionable insights and supporting strategic decision-making processes.

4. New Business Models: AI-powered technologies enable the development of innovative business models, such as personalized recommendations, on-demand services, and predictive maintenance.

5. Job Creation: While AI may replace some jobs, it also creates new job opportunities in AI development, data analysis, and AI system maintenance.

Challenges of AI and Automation

Despite the numerous opportunities, the adoption of AI and automation poses several challenges:

1. Job Displacement: Automation can lead to the displacement of certain jobs, particularly those involving routine and repetitive tasks. This may result in job losses and require workers to acquire new skills to remain employable.

2. Skill Gap: The rapid advancement of AI technologies demands a skilled workforce capable of developing, operating, and maintaining these systems. Addressing the skill gap and providing adequate training and education are crucial to harness the benefits of AI and automation.

3. Ethical Considerations: AI raises ethical concerns regarding privacy, bias, and accountability. Ensuring that AI systems are transparent, fair, and uphold ethical standards is essential for their responsible deployment.

4. Economic Inequality: The unequal distribution of benefits and gains from AI and automation can exacerbate economic inequality. Strategies should be implemented to ensure that the benefits are accessible to all segments of society.

Case Studies

Let's examine two case studies that demonstrate the impact of AI and automation on economic growth:

1. Healthcare: AI-powered systems can analyze medical images, diagnose diseases, and recommend treatment plans. This improves the accuracy and efficiency of diagnoses, reduces healthcare costs, and enhances patient outcomes.

2. Logistics and Transportation: AI algorithms optimize route planning, freight management, and vehicle scheduling, leading to reduced transportation costs, improved delivery times, and enhanced supply chain efficiency.

These case studies highlight the potential of AI and automation to drive transformative changes and deliver significant economic benefits.

Conclusion

AI and automation have the potential to revolutionize various sectors, contributing to economic growth, productivity enhancement, and innovation. However, their adoption also presents challenges such as job displacement and ethical considerations. To harness the benefits of AI and automation while addressing these challenges, it is necessary to foster a supportive environment, ensure inclusive economic growth, and invest in education and skills development. With proper regulations and responsible deployment, AI and automation can play a vital role in shaping the future of economic growth.

Understanding the importance of renewable energy technologies for sustainable development

Renewable energy technologies play a crucial role in achieving sustainable development goals. With the growing concern over climate change and the limited availability of fossil fuels, renewable energy has emerged as a viable and clean alternative. In this section, we will explore the importance of renewable energy technologies in promoting sustainable development and discuss their potential benefits.

Environmental Benefits

One of the key advantages of renewable energy technologies is their minimal impact on the environment. Unlike fossil fuels, renewable energy sources such as solar, wind, and hydropower do not release harmful greenhouse gases into the atmosphere. This contributes to the reduction of carbon emissions and helps mitigate the effects of climate change. Additionally, renewable energy technologies have a lower water footprint compared to conventional power generation methods, reducing water scarcity concerns.

Energy Security

Renewable energy technologies offer greater energy security compared to fossil fuels. Unlike finite fossil fuel reserves, renewable energy sources are inexhaustible and widely available. This reduces dependence on imported energy resources and enhances a country's energy independence. Furthermore, renewable energy systems are often decentralized, allowing for distributed power generation and reducing vulnerability to disruptions in the centralized power grid.

Economic Growth

Investing in renewable energy technologies can stimulate economic growth. The renewable energy sector has the potential to create a significant number of jobs, both in the installation and maintenance of renewable energy systems and in the manufacturing of related components. This can boost local economies and provide new employment opportunities. Moreover, renewable energy technologies have shown potential for attracting foreign direct investment and driving innovation in related industries.

Social Benefits

Renewable energy technologies can bring about various social benefits. Access to affordable and clean energy improves the quality of life and well-being of individuals, particularly in underserved communities. Renewable energy projects can also provide electricity to remote areas that are not connected to the main power grid, promoting social inclusivity. Additionally, the development of renewable energy infrastructure often involves community engagement, empowering local communities and fostering social cohesion.

Challenges and Considerations

While renewable energy technologies offer numerous benefits, there are also challenges and considerations that need to be addressed. Some of the key challenges include:

- **Intermittency**: Renewable energy sources such as solar and wind are intermittent, depending on weather conditions. This poses challenges for grid integration and requires the development of energy storage solutions.
- **Infrastructure and Grid Upgrades**: Shifting to renewable energy requires significant infrastructure and grid upgrades to accommodate increased electricity generation from decentralized sources.
- **Cost and Affordability**: Although the cost of renewable energy technologies has been declining over the years, they still require substantial initial investments. Ensuring affordability is crucial for widespread adoption.
- **Policy and Regulatory Frameworks**: Sustained government support and favorable policy frameworks are essential for encouraging investments in renewable energy technologies.

Case Study: Germany's Energiewende

Germany's Energiewende, or "energy transition," provides a real-world example of a country embracing renewable energy technologies for sustainable development. The Energiewende aims to achieve a low-carbon energy system by increasing the share of renewables in electricity generation, improving energy efficiency, and reducing greenhouse gas emissions.

The success of Germany's Energiewende can be attributed to a combination of factors, including:

- **Policy Support:** Germany has implemented a range of supportive policies and incentives to promote renewable energy, including feed-in tariffs, grants, and tax benefits.

- **Distributed Generation:** The Energiewende has encouraged the development of small-scale renewable energy systems, leading to a decentralized power generation model and increased regional self-sufficiency.

- **Research and Development:** Germany has invested in research and development to drive innovation in renewable energy technologies, fostering a culture of technological advancement.

The Energiewende has not been without challenges, such as grid integration issues and the need for additional storage infrastructure. However, it serves as an inspiring example of how a country can transition to a more sustainable energy future.

Conclusion

Renewable energy technologies are essential for sustainable development. They offer environmental benefits, enhance energy security, stimulate economic growth, and provide social benefits. However, addressing challenges related to intermittency, infrastructure, cost, and policy frameworks is crucial for widespread adoption. Case studies like Germany's Energiewende provide valuable insights into successful implementation strategies. By embracing renewable energy technologies, we can create a more sustainable and resilient future.

Discussing the challenges of technological disruption and the need for adaptation

Technological disruption refers to the rapid and significant changes brought about by the introduction of new technologies. These disruptions can have both positive and negative impacts on various aspects of society, including the economy, workforce, and individual businesses. In this section, we will explore the challenges associated with technological disruption and the need for adaptation.

1. **Job displacement and retraining**: One of the major challenges of technological disruption is the potential displacement of jobs. As technologies such as automation and artificial intelligence advance, certain job sectors may become obsolete, leading to unemployment for many individuals. To address this challenge, it is crucial to provide retraining programs and resources to enable workers to acquire new skills and transition to emerging job opportunities.

2. **Technological skills gap**: Technological disruption often requires individuals to possess new and advanced skill sets. However, there is often a gap between the skills demanded by emerging technologies and the skills possessed by the current workforce. This creates a challenge in terms of providing adequate training and education to bridge this gap and ensure a skilled workforce that can effectively adapt to new technologies.

3. **Inequality and digital divide**: Technological disruption has the potential to exacerbate existing social and economic inequalities. Access to technology and the ability to leverage it for economic opportunities is not evenly distributed. This digital divide creates a challenge in ensuring that all individuals and communities have equal access to technological advancements and the benefits they bring. Efforts must be made to bridge this divide to promote inclusive and equitable development.

4. **Cybersecurity threats**: With the increasing reliance on technology, cybersecurity threats become a significant challenge. As new technologies emerge, so do new vulnerabilities and risks. It is essential to develop robust cybersecurity measures and policies to protect individuals, businesses, and critical infrastructure from cyber-attacks. This requires continuous innovation and adaptation to stay one step ahead of cyber threats.

5. **Ethical implications**: Technological disruption raises ethical concerns that need to be addressed. Emerging technologies such as artificial intelligence and automation bring ethical dilemmas related to privacy, algorithmic bias, job displacement, and the potential misuse of technology. It is crucial to have ethical frameworks and regulations to guide the responsible development and deployment of new technologies.

6. **Resistance to change**: Technological disruption often faces resistance from individuals and organizations comfortable with the status quo. This resistance can slow down the adoption of new technologies and hinder adaptation. Overcoming this challenge requires effective communication, education, and leadership to foster a culture of innovation and openness to change.

7. **Lack of regulatory frameworks**: Rapid technological advancements sometimes outpace the development of necessary regulatory frameworks. This

poses a challenge in terms of ensuring the responsible and ethical use of technologies. Governments and policymakers need to proactively adapt and develop appropriate regulatory frameworks to mitigate risks and maximize the benefits of technological disruption.

8. **Financial implications:** Technological disruption often requires significant investments in research, development, and infrastructure. This can pose a financial challenge, especially for smaller businesses and developing economies. Strategies such as public-private partnerships, government support programs, and access to venture capital can help mitigate financial barriers and promote technological adaptation.

In order to address these challenges, individuals, organizations, and governments must embrace the need for adaptation. This includes:

- Prioritizing investment in education and skills development to ensure individuals have the capabilities to adapt to rapidly changing technological landscapes. - Encouraging a culture of lifelong learning and continuous skill upgrading to foster adaptability and resilience in the face of technological disruptions. - Promoting collaboration and knowledge sharing between individuals, businesses, and governments to facilitate the rapid dissemination of technological advancements. - Supporting research and development initiatives to drive innovation and facilitate the development of new technologies. - Establishing agile regulatory frameworks that balance innovation and consumer protection to enable the responsible and ethical use of technologies. - Nurturing an ecosystem of entrepreneurship and innovation that supports the development and adoption of new technologies.

Overall, technological disruption presents both challenges and opportunities. By recognizing the challenges and embracing the need for adaptation, individuals and societies can navigate these disruptions more effectively and harness the transformative power of technology for economic growth and societal development.

Examining the role of government policies in promoting technological innovation

In today's globalized and interconnected world, technological innovation plays a vital role in driving economic growth and development. Governments have a critical responsibility to create an environment that fosters and promotes technological innovation. This section will explore the various ways in which government policies can influence and support technological innovation.

Understanding the importance of government policies

Government policies have a significant impact on the pace and direction of technological innovation. The right set of policies can create incentives for firms, entrepreneurs, and researchers to invest in research and development (R&D) activities, leading to the discovery of new technologies and the adoption of innovative practices. Moreover, policies can also help in overcoming market failures and addressing societal challenges through targeted investments and regulations.

Examining policy approaches for promoting technological innovation

There are several policy approaches that governments can adopt to promote technological innovation:

1. **Research and Development (R&D) funding:** Governments can play a crucial role in funding R&D activities, especially in sectors with long-term payoffs and high risks. By providing grants, subsidies, and tax incentives, governments can encourage firms to invest in technological innovation. Additionally, public research institutes and universities can receive funding for basic research, which often acts as a foundation for technological advancements.

2. **Intellectual Property Rights (IPR) protection:** Intellectual property rights play a critical role in incentivizing innovation by ensuring that innovators can benefit from their creations. Governments can establish strong legal frameworks for patents, copyrights, trademarks, and trade secrets to protect intellectual property. These protections provide an incentive for firms and individuals to invest in R&D activities without the fear of imitation or theft.

3. **Public procurement policies:** Governments can leverage their purchasing power to drive technological innovation. By directing their procurement policies towards innovative products and services, governments can create demand and encourage firms to invest in R&D. Public procurement policies can also support the diffusion of innovative technologies by providing a market for early-stage innovations.

4. **Regulatory frameworks:** Regulations can either hinder or facilitate technological innovation. Governments can create regulatory frameworks that promote innovation by reducing entry barriers, streamlining approval processes, and ensuring fair competition. Additionally, regulations can also address risks

associated with new technologies, such as privacy concerns or environmental impacts.

5. **Collaboration and knowledge sharing:** To foster technological innovation, governments can promote collaboration between different stakeholders, including firms, academia, research institutes, and NGOs. By establishing innovation clusters, incubators, and technology parks, governments can create an environment where knowledge sharing, networking, and collaboration thrive. Such collaborations can lead to the development of new technologies and the transfer of knowledge across sectors.

Analyzing successful case studies of government policies

Several countries have successfully implemented government policies to promote technological innovation. Let's examine two such case studies:

Case Study 1: South Korea's "Creative Economy" policy South Korea's government implemented the "Creative Economy" policy, which aimed to foster technological innovation and entrepreneurship. The policy focused on promoting collaboration between industries, academia, and the government. It provided financial support for startups and encouraged international collaborations. As a result, South Korea witnessed significant advancements in sectors like information technology, biotechnology, and robotics, and became a global leader in innovation.

Case Study 2: Israel's investment in R&D Israel's government has consistently invested in R&D activities, contributing to its status as a "Start-up Nation." The government provides grants and tax incentives to startups and established companies for R&D investments. Additionally, Israel has a strong network of incubators and venture capital firms that support technological innovation. As a result, Israel has become a hub for technological advancements, particularly in sectors like cybersecurity, agriculture, and healthcare.

Understanding the challenges and opportunities

While government policies play a vital role in promoting technological innovation, there are also challenges that need to be addressed:

1. **Balancing short-term and long-term goals:** Government policies should strike a balance between short-term economic objectives and long-term technological advancements. It can be challenging to make investments in R&D activities that may not yield immediate results but are crucial for future innovation.

2. **Ensuring inclusivity and equitable access:** Government policies should ensure that technological innovation benefits all segments of society and does not exacerbate existing inequalities. Efforts should be made to bridge the digital divide and provide equal access to technological innovations, especially in marginalized communities.

3. **Anticipating and managing risks:** Government policies must consider and address the risks associated with technological innovation. This includes potential negative impacts on employment, privacy, and the environment. Regulations and safeguards should be in place to mitigate these risks and ensure responsible innovation.

Conclusion

Government policies play a crucial role in promoting technological innovation. Through funding R&D activities, protecting intellectual property rights, creating supportive regulatory frameworks, and promoting collaboration, governments can foster an environment that encourages technological advancements. Successful case studies from countries like South Korea and Israel serve as examples of effective policy approaches. However, policymakers must also address challenges such as balancing short-term goals with long-term innovation, ensuring inclusivity, and managing risks. By adopting comprehensive and forward-thinking policies, governments can provide the necessary support for technological innovation and drive economic growth.

Technological Infrastructure and Economic Development

Understanding the concept of technological infrastructure

Defining Technological Infrastructure and Its Components

Technological infrastructure refers to the underlying framework of physical and virtual components that enable the functioning and development of technology-based systems. It encompasses a wide range of interconnected elements that support the transmission, processing, and storage of information, as well as the delivery of goods and services. In this section, we will explore the various components that constitute technological infrastructure and their significance in driving economic development.

    1. **Information and Communication Technologies (ICT)**: ICT refers to the hardware, software, networks, and services used for the collection, storage, processing, and dissemination of information. Key components of ICT include computers, servers, routers, switches, data centers, telecommunications networks, and the internet. ICT infrastructure facilitates communication, collaboration, and knowledge transfer, enabling individuals, organizations, and societies to access and utilize information effectively.

    2. **Transportation Infrastructure:** Transportation infrastructure comprises the physical systems and facilities necessary for the movement of people, goods, and services. It includes roads, railways, airports, seaports, canals, and pipelines. An efficient transportation system is essential for trade activities, supply chain management, and economic integration. It enables the timely and cost-effective movement of goods and facilitates mobility, connectivity, and accessibility.

    3. **Energy Infrastructure:** Energy infrastructure involves the generation,

transmission, and distribution of energy resources required for various economic activities. It encompasses power plants, electrical grids, fuel storage facilities, pipelines, and renewable energy installations. Reliable and affordable energy supply is crucial for industrial production, transportation, and the provision of essential services. Advances in energy infrastructure contribute to sustainable development by promoting cleaner and more efficient energy sources.

4. **Digital Infrastructure:** Digital infrastructure comprises the hardware, software, and networks that facilitate digital connectivity and access to online services. It includes broadband networks, mobile communication towers, cloud computing systems, and data centers. Digital infrastructure is fundamental for the growth of digital economies, allowing individuals and businesses to engage in e-commerce, online communication, and remote work. It also enables the development of smart cities, digital government services, and innovative applications.

5. **Telecommunications Infrastructure:** Telecommunications infrastructure refers to the systems and networks that support the transmission of voice, data, and multimedia content over long distances. It includes telecommunication towers, telephone lines, fiber optic cables, satellites, and communication protocols. A robust telecommunications infrastructure is essential for connectivity, information exchange, and access to telecommunication services such as voice calls, internet access, and broadcasting.

6. **Research and Development (R&D) Infrastructure:** R&D infrastructure encompasses the facilities, equipment, and resources dedicated to scientific research, technological innovation, and development activities. It includes laboratories, research centers, testing facilities, and innovation hubs. R&D infrastructure plays a crucial role in advancing knowledge, fostering technological breakthroughs, and promoting innovation in various sectors of the economy. It provides an environment conducive to collaboration, experimentation, and the translation of research into practical applications.

7. **Financial Infrastructure:** Financial infrastructure comprises the systems and institutions that facilitate financial transactions, investments, and risk management. It includes banks, stock exchanges, payment systems, credit rating agencies, and insurance companies. A well-developed financial infrastructure is essential for mobilizing capital, allocating resources efficiently, and promoting economic growth. It enables individuals and businesses to access financial services, manage risks, and facilitate investments.

8. **Social Infrastructure:** Social infrastructure involves the physical and organizational structures that support social services and community well-being. It includes schools, hospitals, libraries, community centers, and public spaces. Social

infrastructure contributes to human development, education, healthcare, and cultural activities, fostering social cohesion and quality of life. It creates an environment that nurtures creativity, innovation, and social interaction.

9. **Regulatory Infrastructure:** Regulatory infrastructure comprises the laws, regulations, standards, and policies governing the use, development, and operation of technology-based systems. It includes intellectual property laws, data protection regulations, cybersecurity standards, and competition policies. Regulatory infrastructure ensures the protection of intellectual property rights, promotes fair competition, and addresses ethical and legal considerations associated with technology adoption and deployment.

10. **Environmental Infrastructure:** Environmental infrastructure involves the systems and practices that contribute to environmental sustainability and resource management. It includes waste management facilities, water treatment plants, renewable energy installations, and pollution control measures. Environmental infrastructure aims to minimize the impact of economic activities on the environment, promoting sustainable development and the preservation of natural resources.

It is important to note that these components of technological infrastructure are interconnected and interdependent. The development and integration of these components require effective planning, investment, and collaboration between public and private entities. Furthermore, the availability and quality of technological infrastructure influence the capacity of economies to adopt and leverage technology effectively. Policies promoting the development, maintenance, and accessibility of technological infrastructure are crucial for fostering economic development, innovation, and sustainable growth.

Example: A comprehensive technological infrastructure in a country can lead to the expansion of e-commerce. For instance, a well-developed ICT infrastructure with access to high-speed internet and secure payment systems enables businesses and consumers to engage in online transactions. The transportation infrastructure ensures the efficient delivery of goods, while the logistics systems track and manage online orders. Robust energy infrastructure ensures a consistent power supply for running e-commerce platforms and data centers. Digital infrastructure facilitates the development of user-friendly websites and mobile applications. Overall, a strong technological infrastructure enhances the growth of e-commerce, enabling businesses to reach a wider customer base and promoting economic development.

Resource: "Technology Infrastructure and Economic Growth" by Diego A. Comin and Marti Mestieri.

Exercise: Identify three components of technological infrastructure (besides those mentioned above) that are critical for the growth of a digital economy.

Explain why these components are important and provide examples of how they contribute to the development of digital economies.

Examining the importance of infrastructure for economic development

Infrastructure plays a crucial role in the economic development of a country. It refers to the basic physical and organizational structures and facilities needed for the operation of a society or enterprise. In the context of economic development, infrastructure includes various sectors such as transportation, telecommunications, energy, and digital connectivity.

The importance of infrastructure for economic development can be understood by its impact on productivity, competitiveness, and overall economic growth. Infrastructure provides the necessary framework for economic activities to take place efficiently and effectively. Let's explore some key reasons why infrastructure is crucial for economic development:

Enhancing productivity

Infrastructure, such as well-developed transportation networks and efficient logistics systems, facilitates the movement of goods and services. This enables businesses to transport their inputs and outputs more quickly and at lower costs. Improved connectivity through roads, railways, and ports reduces transportation bottlenecks, lowers transaction costs, and enables firms to access a wider market. As a result, productivity improves, leading to increased competitiveness and economic growth.

Enabling trade and integration

Infrastructure acts as a bridge that connects regions and countries, enabling trade and integration. Efficient transportation infrastructure facilitates the movement of goods across borders, making it easier for businesses to participate in international trade. This leads to increased export opportunities, foreign investment, and economic diversification. Additionally, infrastructure can promote regional integration by enhancing connectivity and fostering economic cooperation between neighboring countries.

Attracting investment and fostering entrepreneurship

Investors are attracted to countries with well-developed infrastructure as it provides a conducive environment for business operations. Infrastructure such as reliable energy supply, modern telecommunications networks, and digital connectivity are essential for attracting foreign direct investment. Additionally, infrastructure investments create opportunities for entrepreneurship and job creation, as businesses can leverage the available resources and facilities to innovate, expand, and create new ventures.

Promoting social inclusivity and equitable growth

Infrastructure development plays a crucial role in promoting social inclusivity and equitable growth. Access to basic infrastructure, such as clean water, sanitation, and electricity, improves the quality of life for the population, especially in rural and underserved areas. It enables equal access to education, healthcare, and other essential services, reducing inequality and poverty. Furthermore, investment in infrastructure can create employment opportunities, particularly in labor-intensive construction sectors, contributing to job creation and income generation.

Enabling sustainable development

Infrastructure also plays a vital role in enabling sustainable development. Developing sustainable infrastructure, such as renewable energy systems and energy-efficient buildings, reduces dependency on fossil fuels, mitigates climate change impacts, and promotes environmental sustainability. Additionally, investing in resilient infrastructure and disaster management systems helps countries prepare for and respond to natural disasters, reducing the economic and social costs associated with such events.

Challenges and considerations

While infrastructure development offers immense opportunities for economic development, it is not without challenges and considerations. Some of the key challenges include financing constraints, inadequate planning, policy and regulatory hurdles, and environmental sustainability issues. It is essential to address these challenges and adopt a holistic and sustainable approach to infrastructure development.

In conclusion, infrastructure is a critical component of economic development. It enhances productivity, facilitates trade and integration, attracts investment, promotes social inclusivity, and enables sustainable development. Governments and policymakers need to prioritize infrastructure investments and adopt comprehensive strategies that address the challenges and considerations associated with infrastructure development. By doing so, countries can unlock the full potential of their economies and drive long-term sustainable growth.

Analyzing the impact of technological infrastructure on productivity

Technological infrastructure plays a crucial role in driving economic development and productivity growth. In this section, we will explore the relationship between technological infrastructure and productivity, discussing how the availability and quality of infrastructure can affect the efficiency and output of businesses and industries.

Understanding technological infrastructure

Technological infrastructure refers to the physical and organizational structures that support the functioning of technology and enable its effective use. It encompasses various components such as telecommunications networks, transportation systems, energy grids, and digital infrastructure. These components provide the necessary foundation for the adoption, diffusion, and utilization of technology in an economy.

The importance of infrastructure for economic development

Infrastructure is considered a critical determinant of economic development and productivity. It facilitates the smooth flow of goods, services, information, and people, reducing transaction costs and enabling efficient resource allocation. A well-developed technological infrastructure creates an enabling environment for businesses to operate and thrive, attracting investments, promoting innovation, and driving productivity gains.

Analyzing the impact of technological infrastructure on productivity

1. Enhanced connectivity: Technological infrastructure, particularly telecommunications networks, improves connectivity between firms, individuals, and markets. This enhanced connectivity enables seamless communication, facilitates access to information, and promotes collaboration and knowledge-sharing. As a result, businesses can operate more efficiently, make

faster and informed decisions, and respond quickly to market changes, ultimately boosting productivity.

2. **Efficient transportation systems:** Infrastructure investments in transportation networks, such as roads, railways, and ports, have a direct impact on the movement of goods and services. Efficient transportation systems reduce transportation costs, shorten lead times, and expand market access. This leads to improved supply chain management, streamlined logistics, and increased trade, all of which contribute to higher productivity for businesses.

3. **Reliable energy supply:** A reliable and affordable energy supply is crucial for sustained economic activity and industrial production. Technological infrastructure for energy, including power generation, transmission, and distribution systems, ensures a stable energy supply for businesses. Reliable energy infrastructure reduces downtime, avoids disruptions in operations, and enables firms to operate at full capacity, thereby enhancing productivity.

4. **Digital infrastructure and connectivity:** In today's digital age, digital infrastructure, including high-speed internet connectivity and data centers, has become indispensable for businesses. A robust digital infrastructure allows firms to leverage digital technologies, access online markets, and utilize cloud computing and big data analytics. These digital capabilities enhance productivity by enabling businesses to automate processes, reduce manual work, and leverage data-driven insights for decision-making.

5. **Supporting innovation and technology adoption:** Technological infrastructure provides the necessary support for innovation and technology adoption in businesses. Well-developed infrastructure facilities, such as research and development centers, technology parks, and incubators, create an ecosystem conducive to technological advancements. Businesses can leverage such infrastructure to collaborate with research institutions, access cutting-edge technologies, and develop innovative products and services, all of which drive productivity growth.

Case study: Impact of technological infrastructure on India's productivity

An example of the impact of technological infrastructure on productivity can be seen in India. Over the past two decades, India has significantly invested in its technological infrastructure, particularly in telecommunications and digital connectivity. This investment has led to a significant increase in internet penetration, mobile phone usage, and digital connectivity across the country.

As a result, businesses in India have been able to leverage digital technologies to automate processes, improve supply chain management, and reach new markets. The availability of high-speed internet and digital platforms has enabled small and medium-sized enterprises to connect with customers, access e-commerce platforms, and expand their customer base, ultimately driving productivity gains.

Moreover, the expansion of digital infrastructure has also facilitated the growth of the IT and IT-enabled services sector in India. The country has become a major global hub for outsourcing and offshoring services, attracting multinational companies and generating employment opportunities. These developments have contributed to the overall productivity growth of India's economy.

Challenges and opportunities in developing technological infrastructure

While technological infrastructure offers immense opportunities for productivity growth, there are several challenges that need to be addressed. Some of these challenges include:

1. **Investment requirements:** Developing and maintaining technological infrastructure require significant investments, which may pose financial challenges, particularly for developing countries. Governments and private sector entities need to collaborate to mobilize resources and allocate funds strategically to build and upgrade infrastructure.

2. **Technological advancements:** With rapid technological advancements, infrastructure needs to keep pace with changing requirements. Upgrading and adapting infrastructure to support emerging technologies can be a challenge, requiring careful planning and investment in research and development.

3. **Regional disparities:** Ensuring equitable distribution of technological infrastructure across regions is crucial to avoid regional disparities and promote inclusive growth. Governments should prioritize investments in underserved areas to bridge the digital divide and ensure equal access to infrastructure facilities.

4. **Regulatory environment:** Creating a conducive regulatory environment is essential for the development of technological infrastructure. Clear regulations, supportive policies, and fair competition frameworks can attract private investments and foster innovation in the infrastructure sector.

Despite these challenges, technological infrastructure presents significant opportunities for countries to enhance productivity and drive economic growth. Strategic investments in infrastructure, coupled with effective policies and public-private partnerships, can unlock the potential for productivity gains and contribute to sustainable development.

Summary

In this section, we explored the impact of technological infrastructure on productivity. We discussed how technological infrastructure, encompassing various components such as telecommunications networks, transportation systems, energy grids, and digital infrastructure, plays a critical role in promoting economic development and driving productivity growth. We analyzed the specific ways in which infrastructure improves connectivity, enhances transportation efficiency, ensures reliable energy supply, and supports innovation. We also highlighted the case study of India, demonstrating how investments in technological infrastructure have contributed to productivity gains. Additionally, we identified challenges and opportunities associated with developing technological infrastructure and emphasized the importance of strategic planning and investment in creating an enabling environment for sustainable and inclusive growth.

Discussing the role of telecommunications infrastructure in connecting economies

Telecommunications infrastructure plays a crucial role in connecting economies and enabling economic development. In this section, we will discuss the importance of telecommunications infrastructure, its components, and its impact on economic growth and connectivity.

Defining telecommunications infrastructure and its components

Telecommunications infrastructure refers to the physical and virtual infrastructure that enables the transmission of information over long distances. It includes a wide range of components that work together to facilitate the exchange of data and communication. These components include:

- **Networks:** Telecommunications networks form the backbone of the infrastructure. They consist of a system of interconnected devices, such as computers, routers, switches, and servers, that allow the transmission of data through cables, fiber optic lines, or wireless connections.

- **Connectivity:** The availability of high-speed internet connectivity is essential for telecommunications infrastructure. This includes broadband connections, both fixed and mobile, that enable users to access and transfer data efficiently.

- **Satellite Systems:** Satellite systems are an integral part of the infrastructure, especially in remote areas where laying down traditional cables is not feasible. Satellites facilitate long-distance communication and enable global connectivity.

- **Data Centers:** Data centers are large facilities that house computer systems, data storage, and networking equipment. They provide the necessary infrastructure for storing and processing vast amounts of data.

- **Software and Applications:** Telecommunications infrastructure relies on various software and applications to manage and control the flow of data. These include protocols, encryption algorithms, and communication applications.

Examining the importance of telecommunications infrastructure for economic development

Telecommunications infrastructure is a critical enabler of economic development for several reasons:

1. **Facilitating Communication:** A well-developed telecommunications infrastructure allows businesses, governments, and individuals to communicate and exchange information efficiently. This enhances collaboration, decision-making, and coordination, leading to increased productivity and innovation.

2. **Enabling Access to Information:** Telecommunications infrastructure provides access to a vast array of information and knowledge available on the internet. This empowers individuals and businesses with valuable resources for research, education, and market intelligence.

3. **Supporting E-commerce:** The growth of e-commerce relies heavily on telecommunications infrastructure. With high-speed internet connectivity, businesses can engage in online transactions, expand their customer base, and reach global markets. This opens up new opportunities for economic growth and entrepreneurship.

4. **Fostering Collaboration and Networking:** Telecommunications infrastructure enables businesses and entrepreneurs to connect with partners, suppliers, and customers worldwide. This fosters collaboration, knowledge sharing, and networking, which are essential for innovation and market expansion.

5. **Driving Digital Transformation:** In today's digital age, a robust telecommunications infrastructure is crucial for digital transformation. It enables the adoption of advanced technologies such as cloud computing, data analytics, the Internet of Things (IoT), and artificial intelligence (AI). These technologies drive efficiency, productivity, and competitiveness across various sectors.

Analyzing the impact of telecommunications infrastructure on connectivity

Telecommunications infrastructure plays a pivotal role in connecting economies by improving connectivity at various levels:

1. **National Connectivity:** A well-developed telecommunications infrastructure ensures connectivity within a country, linking urban and rural areas. This helps bridge the digital divide and promotes inclusive economic growth. Individuals and businesses can access services, information, and markets regardless of their geographical location.

2. **International Connectivity:** Telecommunications infrastructure enables countries to connect with the global economy. It provides the means for international communication, trade, and collaboration. Advanced networks and undersea cables facilitate the exchange of data and enable seamless international connectivity.

3. **Business Connectivity:** Telecommunications infrastructure is crucial for businesses to connect with their partners, suppliers, and customers. It allows for smooth communication, enables efficient supply chain management, and supports global operations. This connectivity enhances business competitiveness and drives economic growth.

4. **Individual Connectivity:** Telecommunications infrastructure enables individuals to connect with their social networks, access information, and participate in the digital economy. It empowers them with opportunities for education, employment, and entrepreneurship, thereby improving overall societal well-being.

Understanding the challenges and opportunities in developing telecommunications infrastructure

Developing robust telecommunications infrastructure can be challenging due to various factors:

- **Investment Requirements:** Building and maintaining telecommunications infrastructure requires significant investments in hardware, software, and network expansion. The high capital costs may pose challenges, especially in developing economies with limited resources.

- **Technological Advancements:** The rapid pace of technological advancements necessitates continuous upgrades and innovation in telecommunications infrastructure. Keeping up with the evolving technologies and standards can be demanding and requires substantial resources.

- **Regulatory Framework:** The development of telecommunications infrastructure is closely tied to regulatory policies. Effective regulation is essential to ensure fair competition, consumer protection, and investment incentives. However, striking the right balance between regulation and market freedom can be complex.

- **Geographical Challenges:** Countries with vast territories or challenging geographical landscapes may face difficulties in extending telecommunications infrastructure to remote and underserved areas. Overcoming these challenges requires innovative solutions such as satellite systems or wireless technologies.

Despite these challenges, developing telecommunications infrastructure also presents significant opportunities:

- **Economic Growth and Job Creation:** Developing robust telecommunications infrastructure can stimulate economic growth by attracting investments, promoting entrepreneurship, and creating job opportunities in the telecommunications sector and related industries.

- **Digital Inclusion and Empowerment:** Extending telecommunications infrastructure to underserved areas can bridge the digital divide and promote digital inclusion. It empowers marginalized communities by providing access to information, education, healthcare, and financial services.

- **Innovation and Technological Advancement:** Developing advanced telecommunications infrastructure facilitates innovation and technological advancement. It provides a platform for the adoption and utilization of cutting-edge technologies, opening up new opportunities for research, development, and entrepreneurship.

- **Sustainable Development:** Telecommunications infrastructure can contribute to sustainable development by enabling smarter cities, efficient transportation systems, and sustainable agriculture practices. It supports environmental conservation efforts by promoting remote working and reducing the need for physical travel.

In conclusion, telecommunications infrastructure plays a critical role in connecting economies and fostering economic development. It facilitates communication, enables access to information and markets, and drives digital transformation. Developing robust and inclusive telecommunications infrastructure presents both challenges and opportunities, but the benefits, such as economic growth, innovation, and digital inclusion, make it a worthy investment for countries and societies.

Exploring the importance of transportation infrastructure for trade and economic growth

Transportation infrastructure plays a crucial role in facilitating trade and promoting economic growth. It provides the physical framework for the movement of goods, services, and people, connecting producers with consumers and enabling the efficient functioning of supply chains. In this section, we will examine the various dimensions of transportation infrastructure and its significance for trade and economic development.

Defining transportation infrastructure

Transportation infrastructure refers to the network of physical structures, systems, and facilities that support the movement of goods, services, and people. It encompasses a wide range of components, including roads, railways, airports, seaports, canals, pipelines, and intermodal terminals. These components collectively form the backbone of a country's transportation system and enable the efficient flow of goods and people between different regions, both domestically and internationally.

Importance for trade

Transportation infrastructure is essential for facilitating trade, as it enables the movement of goods to reach markets efficiently. A well-developed transportation system reduces transportation costs, which directly affects the competitiveness of a country's exports in the global market. Efficient transportation infrastructure allows businesses to access inputs, such as raw materials and components, from different regions at lower costs, enhancing their productivity and competitiveness.

Moreover, transportation infrastructure plays a critical role in supporting international trade. Seaports and airports serve as gateways for global trade, facilitating the movement of goods between countries. Efficient seaport operations enable the quick loading and unloading of cargo, reducing port congestion and waiting times. Similarly, well-connected airports with adequate capacity help expedite the movement of air freight, enabling faster and more reliable international trade.

Impact on economic growth

Transportation infrastructure also has a significant impact on economic growth. By improving connectivity and reducing transportation costs, it enhances market access and fosters regional integration. When businesses can connect to larger markets, they can increase their market share, scale up production, and unlock economies of scale. This, in turn, leads to increased output, job creation, and higher incomes, driving economic growth.

Additionally, transportation infrastructure promotes investment and industrial development. Well-connected regions attract investment, as businesses are more likely to locate in areas with reliable transportation networks that facilitate access to inputs, markets, and labor. The presence of transportation infrastructure also supports the development of industrial clusters, as it enables the efficient movement of goods between different stages of production within a region.

Challenges and opportunities

While transportation infrastructure is crucial for trade and economic growth, it faces several challenges. One of the primary challenges is ensuring adequate investment in infrastructure development and maintenance. Developing and maintaining transportation infrastructure requires substantial financial resources, and governments often face budgetary constraints. Finding innovative financing mechanisms, such as public-private partnerships and user fees, can help address this challenge.

Another challenge is ensuring the sustainability of transportation infrastructure. Infrastructure development can have adverse environmental impacts, such as increased carbon emissions and habitat fragmentation. Implementing sustainable transportation solutions, such as investing in public transportation and promoting the use of renewable energy sources, can contribute to mitigating these environmental challenges.

Furthermore, transportation infrastructure needs to be resilient to withstand natural disasters and other disruptions. Investing in infrastructure resilience, such as applying climate-proof design standards and incorporating redundancy in key transportation networks, can help minimize the impact of such disruptions on trade and economic activities.

Despite these challenges, transportation infrastructure also presents opportunities for innovation and technological advancements. Smart transportation systems, leveraging technologies such as Internet of Things (IoT) and artificial intelligence, can enhance the efficiency and safety of transportation networks. For example, intelligent traffic management systems can optimize traffic flows and reduce congestion, improving the overall productivity of transportation systems.

Case study: The impact of transportation infrastructure in China

China serves as a compelling case study on the importance of transportation infrastructure for trade and economic growth. Over the past few decades, China has invested heavily in developing its transportation network, including a vast network of highways, high-speed railways, and modern airports. These infrastructure investments have had a transformative effect on China's economy.

Improved transportation infrastructure has facilitated the integration of China's vast domestic market, connecting different regions and enabling the efficient movement of goods. This has contributed to the growth of manufacturing industries, with China emerging as a global manufacturing powerhouse. Additionally, transportation infrastructure has played a crucial role in supporting China's exports, helping the country become a major player in international trade.

The development of transportation infrastructure in China has also created numerous employment opportunities. Infrastructure projects have created jobs during the construction phase, and the improved connectivity has enabled the growth of industries and services, further boosting employment.

However, the rapid expansion of transportation infrastructure in China has also posed challenges. The country faces issues such as congestion, environmental degradation, and the displacement of communities due to infrastructure development. Addressing these challenges requires a comprehensive and sustainable approach to transportation infrastructure planning and development.

Conclusion

Transportation infrastructure is a critical component of trade and economic development. It facilitates the movement of goods, services, and people, connecting producers with consumers and enabling the efficient functioning of supply chains. By reducing transportation costs, promoting market access, and supporting regional integration, transportation infrastructure drives trade and economic growth. However, it also faces challenges, such as the need for adequate investment, sustainability, resilience, and addressing environmental and social impacts. Embracing innovation and technological advancements can unlock opportunities for enhancing the efficiency and effectiveness of transportation infrastructure systems. By addressing these challenges and seizing opportunities, countries can harness the full potential of transportation infrastructure for trade and economic development.

Investigating the significance of energy infrastructure for industrial development

The development of energy infrastructure plays a crucial role in fostering industrial development and economic growth. Energy is a fundamental input in various industries and serves as the lifeblood of economic activities. In this section, we will explore the significance of energy infrastructure for industrial development, discussing its impact on productivity, competitiveness, and sustainability.

Understanding the importance of energy infrastructure

Energy infrastructure refers to the physical facilities, systems, and networks that enable the generation, transmission, and distribution of energy resources. It encompasses power plants, transmission lines, pipelines, storage facilities, and distribution networks. Without adequate and efficient energy infrastructure, industrial development and economic growth would be severely hampered.

Energy infrastructure provides a reliable and uninterrupted supply of energy, which is essential for industrial activities. It ensures the availability of electricity, oil, gas, and other energy sources to power machinery, equipment, and processes in various sectors, including manufacturing, mining, transportation, and agriculture. A robust and well-maintained energy infrastructure supports the functioning of industries, enabling them to operate at full capacity and meet the growing energy demands.

Analyzing the impact of energy infrastructure on productivity

Energy infrastructure has a significant impact on productivity in industrial sectors. Reliable and affordable energy supply is a critical factor in determining the efficiency and output of production processes. Insufficient or unreliable energy infrastructure can lead to disruptions, downtime, and increased costs in industries.

Adequate energy infrastructure allows industries to operate consistently, achieving higher levels of productivity. It enables the use of energy-intensive technologies and machinery, enhancing manufacturing processes and enabling economies of scale. Industries that rely on energy-intensive operations, such as heavy manufacturing or chemical production, greatly benefit from a robust energy infrastructure that can meet their high energy demands.

Moreover, energy infrastructure can facilitate the adoption of advanced technologies and innovative practices in industries. Modern energy systems, such as smart grids, enable efficient energy management and the integration of renewable energy sources. This not only reduces the environmental impact of industrial activities but also improves energy efficiency, leading to higher productivity and cost savings.

Promoting competitiveness through energy infrastructure

Energy infrastructure plays a crucial role in increasing the competitiveness of industries in the global market. Industries require a reliable and cost-effective energy supply to remain competitive and attract investment. A well-developed energy infrastructure can provide a competitive advantage through its impact on production costs, energy security, and sustainability.

By ensuring a stable and affordable energy supply, energy infrastructure reduces production costs for industries. Energy-intensive sectors, such as steel, aluminum, and cement, heavily depend on energy prices. Access to affordable energy allows these industries to remain competitive both domestically and internationally. Additionally, a robust energy infrastructure can minimize energy price volatility, reducing uncertainty for businesses.

Energy infrastructure also enhances energy security, which is crucial for industrial activities. A diversified and well-connected energy infrastructure reduces dependence on a single energy source or supplier, mitigating the risks associated with energy disruptions or geopolitical events. This enhances the resilience of industries, ensuring a continuous energy supply even in challenging situations.

Furthermore, energy infrastructure contributes to the sustainability and environmental performance of industries. The adoption of clean and renewable energy technologies, facilitated by the availability of appropriate infrastructure, enables industries to reduce their carbon footprint and comply with environmental regulations. This enhances their reputation, attracts environmentally conscious customers, and opens up new market opportunities.

Addressing challenges for sustainable energy infrastructure

Developing sustainable energy infrastructure poses several challenges that need to be addressed for the long-term benefit of industrial development. These challenges include financing, policy and regulatory frameworks, technological innovation, and environmental considerations.

Financing sustainable energy infrastructure projects requires significant investment. Governments, international organizations, and private sector entities need to collaborate to mobilize financial resources for the construction and maintenance of energy infrastructure. Innovative financing mechanisms and public-private partnerships can play a crucial role in overcoming the financial barriers associated with large-scale energy projects.

Effective policy and regulatory frameworks are essential for promoting sustainable energy infrastructure. Governments must create an enabling environment by establishing favorable policies, clear regulations, and attractive incentives to encourage investment in energy infrastructure. Regulatory frameworks should support the integration of renewable energy sources, promote energy efficiency, and ensure the reliable and affordable supply of energy.

Technological innovation plays a vital role in the development of sustainable energy infrastructure. Advancements in renewable energy technologies, energy storage systems, and smart grid solutions are essential to enhance the efficiency and reliability of energy infrastructure. Governments, research institutions, and industry stakeholders should collaborate to promote research and development, as well as the deployment of innovative energy solutions.

Environmental considerations are crucial in the development of energy infrastructure. Projects must adhere to sustainable development principles, minimizing environmental impacts and ensuring the conservation of natural resources. Environmental assessments and monitoring systems should be implemented to evaluate the potential risks and impacts of energy infrastructure projects, taking appropriate measures to mitigate them.

Case study: The role of energy infrastructure in China's industrial development

China's rapid industrialization and economic growth have been closely tied to its investment in energy infrastructure. The country has heavily invested in building power plants, transmission lines, and transportation networks to meet the energy demands of its industries. This robust energy infrastructure has played a pivotal role in transforming China into the world's largest manufacturing hub.

China's energy infrastructure investments have supported the growth of energy-intensive industries, such as steel, cement, and chemicals. Reliable and affordable energy supply has allowed these industries to expand their production capacities and compete globally. Furthermore, the country's investments in renewable energy infrastructure, such as wind and solar power, have fostered the development of clean and sustainable industries.

The development of energy infrastructure in China has faced challenges, including environmental concerns and the need to address energy efficiency. However, the government has implemented policies and regulations to promote sustainable energy practices and reduce carbon emissions. China's experience demonstrates the importance of a comprehensive approach to energy infrastructure development, considering both economic and environmental factors.

Key takeaways

- Energy infrastructure is crucial for industrial development, providing a reliable and uninterrupted energy supply for various sectors. - Energy infrastructure significantly impacts productivity by enabling efficient production processes and the adoption of advanced technologies. - Well-developed energy infrastructure enhances the competitiveness of industries through reduced production costs, improved energy security, and sustainability. - Challenges for sustainable energy infrastructure include financing, policy frameworks, technological innovation, and environmental considerations. - Case studies, such as China's energy infrastructure development, provide insights into the role of energy infrastructure in industrial growth.

Summary

In this section, we investigated the significance of energy infrastructure for industrial development. We discussed how energy infrastructure impacts productivity and competitiveness in industries. Additionally, we explored the challenges associated with developing sustainable energy infrastructure and examined a case study of China's energy infrastructure development. A robust and sustainable energy infrastructure is crucial for fostering industrial development and ensuring a sustainable and prosperous economy.

Assessing the role of digital infrastructure in promoting digital economies

In today's interconnected and digital world, having a robust digital infrastructure is crucial for promoting and sustaining digital economies. Digital infrastructure refers to the underlying technological framework that enables the smooth functioning of digital systems, networks, and services. This includes telecommunications networks, internet connectivity, data centers, and other critical components that support the exchange and processing of digital information.

Defining digital infrastructure and its components

Digital infrastructure encompasses a wide range of components that work together to facilitate the flow of information and support digital activities. These components include:

1. **Telecommunications networks:** These networks, including fiber optic cables, satellite systems, and wireless networks, provide the backbone for digital connectivity. They enable the transmission of data, voice, and video signals over long distances, connecting individuals, businesses, and governments worldwide.

2. **Internet connectivity:** High-speed and reliable internet connectivity is essential for accessing digital services and participating in the digital economy. This includes both fixed-line broadband connections and wireless networks, such as 4G and 5G.

3. **Data centers:** Data centers are secure facilities that house computer systems, storage devices, and network infrastructure. They serve as the central hub for storing, managing, and processing digital data, enabling the provision of various online services.

4. **Cloud computing infrastructure:** Cloud computing infrastructure allows users to access computing resources, such as servers, storage, and software, over the internet on a pay-per-use basis. It provides scalability, flexibility, and cost-efficiency for businesses and individuals leveraging digital technologies.

5. **Cybersecurity infrastructure:** With the increasing threats of cyber-attacks and data breaches, a robust cybersecurity infrastructure is essential for protecting digital systems and ensuring the privacy, integrity, and availability of digital assets.

6. **Digital payment systems:** Seamless and secure digital payment systems, including electronic banking, mobile wallets, and digital currencies, play a vital role in enabling online transactions and e-commerce activities.

7. **Data analytics and artificial intelligence (AI) capabilities:** Leveraging data analytics and AI technologies allows organizations to extract valuable insights, automate processes, and drive innovation in various sectors of the digital economy.

Examining the importance of infrastructure for economic development

A robust digital infrastructure is a critical enabler of economic development in the digital age. It provides the foundation for digitalization, innovation, and the creation of new digital business models. Here are some key reasons why digital infrastructure is essential for economic development:

1. **Facilitating digital connectivity:** Digital infrastructure ensures that individuals, businesses, and governments can connect, communicate, and access important digital resources. This connectivity allows for the efficient exchange of information, collaboration, and the delivery of digital services.

2. **Enabling e-commerce and digital trade:** A well-developed digital infrastructure enables the growth of e-commerce and digital trade, which have become significant contributors to economic growth globally. It allows businesses to reach a wider customer base, reduces transaction costs, and fosters international trade.

3. **Promoting innovation and entrepreneurship:** Digital infrastructure provides the necessary platform for innovation and the development of digital startups. It facilitates the adoption of new technologies, supports research and development, and fosters a dynamic digital ecosystem that encourages entrepreneurial activities.

4. **Driving productivity gains:** Digital technologies, supported by a robust digital infrastructure, can significantly enhance productivity across sectors of the economy. Automation, data-driven decision-making, and process

optimization are just a few examples of how digital infrastructure can drive efficiency and productivity gains.

5. **Attracting investment and talent:** Countries with advanced digital infrastructure often attract investment from digital companies and entrepreneurs. A strong digital infrastructure signals a conducive environment for technology-driven businesses and innovation, attracting both domestic and foreign talent.

Analyzing the impact of digital infrastructure on productivity

Digital infrastructure plays a crucial role in boosting productivity by enabling efficiency gains and innovation. Here are some ways digital infrastructure impacts productivity:

1. **Improved connectivity and communication:** Digital infrastructure enables seamless connectivity and communication, allowing for real-time collaboration, faster decision-making, and efficient information exchange. This leads to improved coordination and productivity gains, particularly in remote work environments and global business operations.

2. **Automation and digitization of processes:** Digital infrastructure supports the automation and digitization of manual and paper-based processes. By eliminating manual tasks, businesses can achieve cost savings, reduce errors, and improve the overall efficiency of their operations.

3. **Data-driven decision-making:** Digital infrastructure provides the foundation for collecting, processing, and analyzing vast amounts of data. By leveraging advanced analytics tools and AI technologies, organizations can gain valuable insights from data, which can inform strategic decision-making and drive productivity improvements.

4. **E-commerce and digital platforms:** A well-developed digital infrastructure enables the growth of e-commerce and digital platforms, which have transformed the way companies conduct business. Digital platforms provide new avenues for reaching customers and streamline transactions, leading to increased sales, cost savings, and productivity gains.

5. **Enhanced access to knowledge and resources:** Digital infrastructure facilitates access to a wealth of knowledge, educational resources, and online learning platforms. This enables individuals to acquire new skills and

knowledge, improving their productivity and employability in the digital economy.

Discussing the role of telecommunications infrastructure in connecting economies

Telecommunications infrastructure, a key component of digital infrastructure, plays a vital role in connecting economies and enabling digital transformation. Here are some ways in which telecommunications infrastructure supports economic development:

1. **Improved connectivity and access to information:** Telecommunications networks, including broadband internet connections, enable individuals and businesses to access information, services, and markets. This promotes economic inclusion, facilitates knowledge exchange, and supports entrepreneurship and innovation.

2. **Strengthened global competitiveness:** Countries with advanced telecommunications infrastructure are better positioned to attract investment, promote trade, and compete in the global market. High-speed internet connections and reliable telecommunications networks enable businesses to leverage advanced digital technologies and participate in global value chains.

3. **Promotion of digital entrepreneurship and startups:** Telecommunications infrastructure provides the necessary connectivity for digital entrepreneurs and startups to develop and scale their businesses. Access to high-speed internet connections enables these businesses to access global markets, collaborate with partners, and leverage digital platforms and tools.

4. **Support for remote work and telecommuting:** Robust telecommunications infrastructure facilitates remote work and telecommuting arrangements. This not only provides flexibility for employees but also enables businesses to tap into a wider pool of talent, regardless of geographical location.

5. **Enabling smart cities and digital services:** Telecommunications infrastructure forms the backbone of smart city initiatives and the provision of digital services. From connected sensors to enable efficient resource management to improved public services through digital applications, telecommunications infrastructure is crucial for enhancing the quality of life in urban areas.

Exploring the importance of transportation infrastructure for trade and economic growth

While digital infrastructure is essential, it must work hand in hand with physical infrastructure to enable economic growth. Transportation infrastructure, in particular, plays a critical role in supporting the movement of goods, people, and ideas. Here are some key reasons why transportation infrastructure is vital for trade and economic growth in the digital economy:

1. **Facilitating the movement of goods:** Transportation infrastructure, including roads, railways, ports, and airports, allows for the efficient movement of goods across regions and countries. This is particularly important in the context of global supply chains, where timely delivery is vital for businesses to meet customer demands.

2. **Reducing trade costs:** Well-developed transportation infrastructure helps reduce trade costs by improving logistics efficiency, reducing transportation times, and minimizing disruptions. This encourages trade and enables businesses to access a wider range of inputs and markets, boosting economic growth.

3. **Promoting regional integration:** Efficient transportation infrastructure is a catalyst for regional integration and economic cooperation. It allows for the seamless movement of goods, services, and people across borders, fostering trade relations and promoting economic cooperation between countries.

4. **Enabling tourism and travel:** Adequate transportation infrastructure, including airports, road networks, and public transportation systems, supports tourism and travel. This contributes to economic growth by attracting visitors, creating job opportunities, and generating revenue for local economies.

5. **Connecting remote areas and promoting inclusivity:** Transportation infrastructure plays a crucial role in connecting remote and rural areas, bridging the digital divide, and promoting economic inclusivity. It enables these areas to access markets, education, healthcare, and social services, driving economic development and reducing regional disparities.

Understanding the challenges and opportunities in developing technological infrastructure

Developing and maintaining technological infrastructure poses various challenges and opportunities for governments, businesses, and society as a whole. Some of these challenges and opportunities include:

1. **Funding and investment:** Developing technological infrastructure requires significant funding and investment. Governments and private stakeholders need to allocate resources and develop sustainable financing models to ensure the continuous development and maintenance of digital infrastructure.

2. **Addressing the digital divide:** Ensuring equitable access to digital infrastructure is critical. Efforts should be made to bridge the digital divide by providing access to underserved areas, marginalized communities, and vulnerable populations. This requires addressing affordability, availability, and digital skills gaps.

3. **Cybersecurity and privacy concerns:** As digital infrastructure becomes more pervasive, the risks of cyber threats and privacy breaches increase. Robust cybersecurity measures and privacy frameworks must be put in place to protect digital infrastructure from malicious actors and ensure the security and trust of digital systems.

4. **Regulatory frameworks and policy coordination:** Developing appropriate regulatory frameworks and ensuring policy coordination across sectors is essential. This includes spectrum allocation, competition policies, data protection regulations, and standards for interoperability to promote seamless connectivity and innovation.

5. **Collaboration and partnerships:** Developing technological infrastructure requires collaboration and partnerships between various stakeholders, including governments, private sector entities, civil society organizations, and research institutions. Public-private partnerships and knowledge sharing platforms can leverage resources and expertise to overcome infrastructure development challenges.

6. **Skills development and digital literacy:** Building the necessary skills and digital literacy among individuals is vital. Efforts should be made to provide education and training programs that equip individuals with the skills needed to effectively leverage digital technologies and participate in the digital economy.

Analyzing successful case studies of countries with robust technological infrastructure

Several countries have successfully developed robust technological infrastructure, contributing to their economic competitiveness and digital transformation. Here are a few noteworthy case studies:

1. **South Korea:** South Korea is known for its advanced telecommunications infrastructure and high-speed internet connectivity. The country has made substantial investments in broadband networks, leading to widespread adoption of digital technologies and the growth of its digital economy.

2. **Singapore:** Singapore has established itself as a global technology and innovation hub. With its well-developed digital infrastructure, including high-speed internet connectivity and smart city initiatives, Singapore has attracted multinational companies and startups, driving economic growth and technological advancements.

3. **Estonia:** Estonia is a pioneer in digital governance and e-services. The country has built a robust digital infrastructure that enables citizens to access government services online, promotes digital entrepreneurship, and supports a thriving e-residency program, attracting businesses from around the world.

4. **Sweden:** Sweden is recognized for its focus on innovation and digital transformation. The country has invested in digital infrastructure, promoting high-speed internet access, and encouraging the adoption of digital technologies in various sectors. This has contributed to the growth of innovative startups and a thriving digital ecosystem.

5. **United Arab Emirates (UAE):** The UAE has made significant investments in digital infrastructure, aiming to become a global leader in technology and innovation. The country has implemented initiatives like Smart Dubai, which leverage digital infrastructure to enhance government services, promote digital entrepreneurship, and foster a culture of innovation.

These case studies highlight the importance of strategic investments, supportive policies, and collaboration between stakeholders in developing technological infrastructure for sustained economic development and digital transformation.

Understanding the challenges and opportunities in developing technological infrastructure

Technological infrastructure plays a crucial role in economic development by providing the foundation for the adoption and implementation of various technologies. It encompasses a wide range of components, including telecommunications networks, transportation systems, energy grids, and digital infrastructure. Developing robust technological infrastructure is essential for enhancing productivity, fostering innovation, enabling connectivity, and promoting overall economic growth. However, there are numerous challenges and opportunities associated with the development of technological infrastructure.

Challenges

1. **Financial constraints:** One of the primary challenges in developing technological infrastructure is the availability of financial resources. Building and maintaining infrastructure require significant investments, which can be hindered by limited public funding or lack of private sector involvement. Governments often face budgetary constraints, and attracting private sector investments can be challenging, especially in less developed regions.

2. **Technological complexities:** Developing technological infrastructure involves dealing with complex systems and technologies. Implementing high-speed telecommunications networks, modern transportation systems, or reliable energy grids requires expertise in various engineering disciplines. Overcoming technical challenges and ensuring the compatibility, scalability, and resilience of infrastructure components can be a daunting task.

3. **Policy and regulatory issues:** The development of technological infrastructure also faces regulatory and policy challenges. Governments need to formulate effective policies and regulations to foster the development and deployment of infrastructure. Balancing the interests of multiple stakeholders, ensuring fair competition, and addressing potential monopolies or anti-competitive behaviors are crucial for successful infrastructure development.

4. **Geographic constraints:** Geographic factors can pose significant challenges in infrastructure development, particularly in remote areas or regions with challenging terrain. Building transportation networks or installing energy infrastructure in mountainous regions or isolated islands can be expensive and technically complex. Overcoming these geographic constraints requires innovative solutions and careful planning.

5. **Environmental considerations:** Developing technological infrastructure should also consider environmental sustainability. Infrastructure projects can have significant environmental impacts, such as deforestation, habitat destruction, or carbon emissions. Striking a balance between economic development and environmental conservation is essential to ensure long-term sustainability.

Opportunities

Despite the challenges, developing technological infrastructure presents several opportunities for economic development and societal progress.

1. **Enhanced connectivity:** Technological infrastructure, such as high-speed telecommunications networks and digital infrastructure, enables connectivity and access to information, improving communication and collaboration. It facilitates e-commerce, online education, telemedicine, and remote working, unlocking new opportunities for economic growth and societal development.

2. **Increased productivity:** Well-developed technological infrastructure enhances productivity by enabling efficient transportation, reliable energy supply, and advanced communication systems. It reduces transaction costs, streamlines supply chains, and automates processes, leading to improved efficiency, reduced waste, and increased output.

3. **Innovation and entrepreneurship:** Technological infrastructure acts as a catalyst for innovation and entrepreneurship. It provides a platform for the development and deployment of new technologies, fostering the growth of startups, and attracting investments. Access to reliable infrastructure encourages experimentation, collaboration, and the emergence of new business models.

4. **Regional development:** Developing technological infrastructure can promote regional development by attracting investments, creating job opportunities, and improving the quality of life. Well-connected regions with robust infrastructure are more likely to attract businesses, industries, and skilled labor, leading to balanced regional development.

5. **Sustainable development:** Technological infrastructure offers opportunities for sustainable development by promoting the adoption of clean technologies. Well-planned energy infrastructure can support the transition to renewable energy sources, reducing greenhouse gas emissions and mitigating climate change. Smart transportation systems can optimize resource utilization, reduce congestion, and minimize environmental impacts.

Examples

To illustrate the challenges and opportunities in developing technological infrastructure, let's consider a few examples:

Example 1: High-speed internet in rural areas: In many rural regions, the lack of high-speed internet access poses a significant challenge for economic development. Limited availability of internet connectivity hampers education, e-commerce, and access to online services. The challenge here lies in bridging the digital divide by investing in broadband infrastructure and implementing policies to promote internet access even in remote areas. Overcoming this challenge can unlock opportunities for rural communities to participate in the digital economy, access online education resources, and establish digital businesses.

Example 2: Renewable energy infrastructure: Developing renewable energy infrastructure presents both challenges and opportunities. The challenge lies in transitioning from fossil fuel-based energy systems to clean and sustainable alternatives. This transition requires significant investments in renewable energy generation, transmission, and storage infrastructure. However, by embracing renewable energy sources and adopting modern grid technologies, countries can reduce dependency on imported fossil fuels, minimize greenhouse gas emissions, and create new job opportunities in the clean energy sector.

Example 3: Smart transportation systems: Developing smart transportation systems can significantly improve mobility, safety, and efficiency. However, the challenge lies in integrating various technologies, such as intelligent traffic management systems, electric vehicle charging infrastructure, and autonomous vehicles. Overcoming technical and regulatory barriers can create opportunities for developing cities to reduce congestion, lower carbon emissions, and enhance public transportation services.

Conclusion

Developing technological infrastructure is essential for economic growth and societal progress. While there are challenges to overcome, such as financial constraints, technological complexities, policy issues, geographic limitations, and environmental considerations, leveraging available opportunities can lead to enhanced connectivity, increased productivity, innovation, regional development, and sustainable growth. Governments, policymakers, and various stakeholders must work together to address the challenges and seize the opportunities, fostering inclusive, resilient, and future-ready technological infrastructure.

Analyzing successful case studies of countries with robust technological infrastructure

In this section, we will analyze several case studies of countries that have successfully developed robust technological infrastructure. These case studies provide valuable insights into the strategies and approaches used by these countries to accelerate their economic development through technology adoption and diffusion.

Case Study 1: South Korea

South Korea is often cited as one of the most technologically advanced countries in the world. The country has made significant investments in building a comprehensive technological infrastructure, which has played a crucial role in its economic development.

One key aspect of South Korea's success is its emphasis on telecommunications infrastructure. The country has developed one of the most advanced and widespread broadband networks, enabling high-speed internet access to almost all citizens. This has facilitated the growth of digital industries and supported the development of innovative technologies.

Additionally, South Korea has invested heavily in research and development (R&D) activities. The government has established public-private partnerships to foster collaboration between academia, industry, and research institutions. These partnerships have led to the development of cutting-edge technologies in areas such as information and communication technology (ICT), semiconductors, and biotechnology.

Furthermore, South Korea has focused on nurturing a skilled workforce through its education system. The country has placed a strong emphasis on science, technology, engineering, and mathematics (STEM) education at all levels. This has contributed to the availability of a highly skilled labor force that can effectively contribute to technological development and innovation.

The success of South Korea's technological infrastructure can be seen in the growth of its high-tech industries, such as electronics, automotive, and telecommunications. These industries have not only propelled the country's economic growth but have also created numerous job opportunities and bolstered its global competitiveness.

Case Study 2: Singapore

Singapore is another exemplary case of a country with a robust technological infrastructure. The country has strategically positioned itself as a global hub for technology and innovation, attracting multinational companies and fostering a thriving startup ecosystem.

One key factor contributing to Singapore's success is its commitment to developing a strong digital infrastructure. The government has invested heavily in building a reliable and pervasive broadband network, enabling seamless connectivity across the nation. This digital infrastructure has been critical in supporting various industries, including finance, logistics, and healthcare.

Furthermore, Singapore has actively promoted research and development activities through partnerships between the public and private sectors. The government has established research institutes and innovation centers to facilitate collaboration between industry and academia. These initiatives have resulted in the emergence of groundbreaking technologies in areas such as smart cities, artificial intelligence, and biomedicine.

Singapore's emphasis on education and talent development has also played a significant role in its technological advancement. The country has prioritized STEM education and offers a range of specialized programs and scholarships to nurture young talents in science and technology. This focus on human capital development has attracted top-tier talent from around the world, strengthening Singapore's innovation ecosystem.

Moreover, Singapore has adopted a proactive approach to attract foreign direct investment (FDI) in technology-related industries. The government provides incentives and support for companies to establish their regional headquarters or research facilities in the country. This has fostered knowledge transfer and created a conducive environment for technology adoption and diffusion.

The case of Singapore illustrates how a small nation can leverage its technological infrastructure to become a global leader in innovation and economic development.

Case Study 3: Estonia

Estonia, a small country in Northern Europe, has made remarkable progress in developing a robust technological infrastructure. The nation has embraced digital transformation and leveraged technology to streamline government services, boost economic growth, and improve the quality of life for its citizens.

One key aspect of Estonia's success is its comprehensive digital infrastructure. The country has implemented a secure and user-friendly digital ID system that allows citizens to access government services, sign documents, and conduct online transactions. This has significantly reduced bureaucratic inefficiencies and enhanced the efficiency of public service delivery.

Additionally, Estonia has invested in building a reliable and high-speed internet network. The country has one of the highest rates of broadband penetration in the world, enabling seamless connectivity for individuals and businesses. This has attracted technology companies and foreign direct investment, driving economic growth and creating job opportunities.

Estonia has also focused on nurturing a culture of innovation and entrepreneurship. The country has implemented policies to support startups and small businesses, including simplified registration procedures, tax incentives, and access to funding. These initiatives have fostered a vibrant startup ecosystem and resulted in the emergence of innovative technologies in areas such as e-governance, e-residency, and cybersecurity.

Furthermore, Estonia has prioritized digital literacy and education. The government has integrated technology into the national curriculum and provides training programs to develop digital skills among citizens. This focus on digital literacy has empowered individuals to fully harness the potential of technology and has contributed to Estonia's digital success.

The case of Estonia highlights the importance of a supportive policy environment, strong digital infrastructure, and a focus on digital literacy in driving technological development and economic growth.

Key Takeaways

The case studies of South Korea, Singapore, and Estonia provide valuable insights into the strategies and approaches employed by countries to develop robust technological infrastructure. Here are some key takeaways from these case studies:

1. Investment in telecommunications infrastructure is crucial for enabling widespread access to high-speed internet and supporting digital industries. 2. Collaboration between academia, industry, and research institutions is essential for fostering technological innovation. 3. Emphasis on STEM education and talent development can contribute to building a skilled workforce capable of driving technological advancement. 4. Developing a supportive policy environment, including incentives for research and development, can attract foreign direct investment and facilitate technology adoption. 5. Digital literacy programs and initiatives are necessary to ensure individuals can fully leverage the benefits of technology.

By studying these successful case studies, policymakers, researchers, and entrepreneurs can gain valuable insights into the factors that contribute to the development of a robust technological infrastructure and use them to inform their own strategies and initiatives.

Discussing the role of public-private partnerships in developing technological infrastructure

Public-private partnerships (PPPs) play a crucial role in the development of technological infrastructure. They are collaborative ventures between government entities and private sector organizations that aim to leverage the strengths of both sectors to achieve common goals. In the context of technological infrastructure, PPPs involve the joint planning, financing, construction, operation, and maintenance of infrastructure projects.

Understanding the need for public-private partnerships

The development of technological infrastructure requires significant investment and expertise. Government entities may face limitations in terms of financial resources, specialized technical knowledge, and operational efficiency. On the other hand, private sector organizations possess capital, technological know-how, and innovative solutions. By partnering together, governments and private sector entities can overcome these limitations and create a more conducive environment for technological development.

PPPs provide a platform for sharing risks, costs, and responsibilities. They enable the pooling of resources, access to cutting-edge technologies, and the utilization of private sector efficiency and innovation. In the context of technological infrastructure, PPPs can lead to the timely and cost-effective development of projects, enhanced service quality, and the delivery of innovative solutions that address the evolving needs of society.

Key benefits of public-private partnerships in developing technological infrastructure

1. **Cost-sharing and efficiency:** PPPs allow the sharing of financial risks between the public and private sectors. The private sector can bring in capital investment and expertise, reducing the burden on the government budget. The involvement of private sector entities also brings in a profit-driven mindset and efficiency to project management, leading to cost savings.

2. **Innovation and technology transfer:** Private sector organizations are at the forefront of technological advancements. Through PPPs, governments can harness the expertise and innovative capabilities of the private sector to leverage state-of-the-art technologies and knowledge transfer. This promotes the development and adoption of modern and efficient technological infrastructure.

3. **Accelerated project timelines:** PPPs can expedite the development of technological infrastructure by leveraging private sector expertise and streamlined decision-making processes. Private sector entities are motivated by profitability and timelines, leading to faster project execution compared to traditional public procurement methods. This is particularly essential in the rapidly evolving technological landscape.

4. **Risk management and long-term sustainability:** PPPs enable risk sharing between the public and private sectors. The private sector assumes certain risks associated with project development, operation, and maintenance. This ensures the financial viability of the project and incentivizes private sector entities to maximize the long-term sustainability of the technological infrastructure.

5. **Service quality and innovation:** PPPs introduce competition and performance-based contracts, driving private sector entities to deliver high-quality services and innovative solutions. This leads to improved service delivery, enhanced user experience, and the provision of technologically advanced infrastructure that caters to the evolving needs of businesses and society.

Challenges and considerations in implementing public-private partnerships

While PPPs offer numerous advantages, their successful implementation requires careful consideration of certain challenges and factors. Some key challenges include:

1. **Risk allocation:** Allocating risks effectively between public and private partners can be complex. Proper risk identification, assessment, and mitigation strategies are essential to minimize uncertainties and ensure project success.

2. **Balancing public interest and private profit:** PPPs must strike a balance between public interest and private profit. Transparency, accountability, and robust governance mechanisms are crucial to safeguarding the public interest, preventing misuse of public funds, and ensuring the delivery of services to all sections of society.

3. **Legal and regulatory frameworks:** Clear legal and regulatory frameworks are essential to govern PPP arrangements. These frameworks should address issues such as contract management, dispute resolution mechanisms, and protection of public interest, while providing a conducive environment for private sector participation.

4. **Financial viability and affordability:** PPPs require careful financial structuring to ensure the long-term financial viability of projects. Balancing affordability for users and the profitability of private sector entities is a critical consideration.

5. **Procurement and oversight mechanisms:** Developing robust procurement processes and oversight mechanisms is crucial to prevent corruption, favoritism, or

undue influence in the selection of private partners and the management of PPP projects.

Case study: The Broadband PPP in Australia

An example of a successful public-private partnership in developing technological infrastructure is the National Broadband Network (NBN) project in Australia. The NBN project aims to provide high-speed broadband access to all Australians, bridging the digital divide and driving economic growth.

The Australian government partnered with private sector companies to design, build, and operate the NBN. Through the PPP model, the government provided funding, ensured regulatory support, and established governance frameworks, while the private sector partners contributed capital investment, technical expertise, and operational efficiencies.

The NBN project has resulted in significant advancements in Australia's technological infrastructure. It has increased access to high-speed internet across the country, enabling improved connectivity, digital services, and opportunities for businesses and individuals. The PPP model allowed for the timely deployment of the network, leveraging private sector efficiencies and expertise.

Conclusion

Public-private partnerships play a critical role in developing technological infrastructure. By harnessing the expertise, resources, and innovation of both the public and private sectors, PPPs facilitate the timely, cost-effective, and sustainable development of projects. However, effective implementation requires careful consideration of risks, regulatory frameworks, financial viability, and public interest. Through successful partnerships, technological infrastructure can be developed to meet the evolving needs of society, driving economic growth and improving the quality of life for individuals and businesses.

Key Technologies for Economic Development

Examining the role of information and communication technologies (ICT) in economic growth

Information and communication technologies (ICT) play a crucial role in driving economic growth in modern societies. The rapid advancements in ICT have revolutionized the way businesses operate and have transformed various sectors of the economy. In this section, we will explore the key contributions of ICT to economic growth, understand the underlying principles of ICT, discuss the challenges and opportunities associated with ICT adoption, and analyze successful case studies of countries leveraging ICT for economic development.

Understanding the principles of information and communication technologies

Information and communication technologies encompass a broad range of tools and systems that enable the creation, storage, processing, and dissemination of information. These technologies include hardware devices (such as computers, smartphones, and servers), software applications, networks, and the internet. The principles of ICT are rooted in the fields of computer science, telecommunications, and information systems.

At the core of ICT is the concept of digitalization, which refers to the conversion of analog data into digital form. Digital data can be easily stored, manipulated, and transmitted using computers and networks, enabling efficient information processing and communication. The advancement of microelectronics, the internet, and multimedia technologies has paved the way for the rapid development and adoption of ICT worldwide.

Contributions of information and communication technologies to economic growth

The contributions of ICT to economic growth are manifold and extend to various aspects of the economy. Here, we highlight some key areas where ICT has made a significant impact:

1. **Productivity enhancement:** ICT has the potential to significantly enhance productivity by automating manual tasks, streamlining processes, and improving communication and collaboration within organizations. For example, the use of enterprise resource planning (ERP) systems enables efficient management of resources, leading to cost savings and increased productivity. Additionally, ICT enables the implementation of advanced manufacturing technologies, such as robotics and automation, which enhance production efficiency.

2. **Innovation facilitation:** ICT serves as a catalyst for innovation by providing the tools and platforms for research and development activities. The availability of vast amounts of data and advanced analytics capabilities

enables companies to gain valuable insights, identify new market opportunities, and develop innovative products and services. Additionally, collaboration and knowledge-sharing platforms facilitated by ICT promote co-innovation and open innovation practices, leading to accelerated technological advancements.

3. **Business transformation:** ICT has transformed business models and enabled the emergence of new industries and markets. E-commerce platforms have revolutionized the retail sector, enabling businesses to reach customers globally and streamline their operations. The gig economy, enabled by ICT platforms, has also created new opportunities for income generation and employment. Moreover, digital marketing and analytics tools have empowered companies to better understand customer preferences and tailor their marketing strategies accordingly.

Challenges and opportunities in ICT adoption

While ICT offers immense opportunities for economic growth, its adoption and implementation are not without challenges. Some key challenges and opportunities associated with ICT adoption are:

1. **Infrastructure requirements:** The widespread adoption of ICT requires robust physical infrastructure, including reliable electricity supply, internet connectivity, and data centers. Developing countries often face challenges in infrastructure development, hindering the effective utilization of ICT for economic growth. However, investments in infrastructure development can create opportunities for economic development and technological leapfrogging.

2. **Digital divide:** The digital divide refers to the gap between individuals, communities, and countries in accessing and utilizing ICT. Socioeconomic factors, education levels, and geographical location can contribute to this divide. Bridging the digital divide is crucial to ensure equal opportunities for all and to leverage the full potential of ICT for economic growth. Initiatives such as digital literacy programs and affordable access to ICT infrastructure can help address this challenge.

3. **Cybersecurity concerns:** The widespread adoption of ICT also opens up vulnerabilities to cyber threats and attacks. Protecting digital assets, securing data, and ensuring the privacy of users are critical challenges in the

digital age. Governments, organizations, and individuals need to invest in cybersecurity measures, develop robust policies, and promote cyber awareness to mitigate these risks.

4. **Skills development:** The rapid pace of technological advancements requires a skilled workforce capable of leveraging ICT effectively. Upskilling and reskilling programs are essential to equip individuals with the necessary digital literacy and technical expertise. Additionally, promoting STEM education and fostering a culture of lifelong learning are crucial for addressing skill gaps and enabling inclusive economic growth.

Case studies: Leveraging ICT for economic development

Several countries have successfully leveraged ICT for economic development. Let's examine two such case studies:

1. **Estonia:** Estonia is often considered a pioneer in ICT adoption and digital governance. The country has implemented various e-services, including e-residency, digital signatures, and online voting, which have streamlined administrative processes and improved efficiency. Estonia's strong ICT infrastructure, combined with a focus on digital literacy and cybersecurity, has attracted foreign investments and fostered the growth of ICT-based startups.

2. **South Korea:** South Korea's focus on ICT development, coupled with extensive investments in infrastructure, has propelled the country to become a global leader in ICT adoption. The government's initiatives, such as the Broadband Convergence Network (BcN) and public e-services, have led to improved connectivity and enhanced efficiency in various sectors. South Korean companies have also capitalized on ICT advancements, notably in the electronics and telecommunications industries.

These case studies highlight the importance of supportive policies, infrastructure development, and digital literacy in harnessing the potential of ICT for economic growth.

Summary

In this section, we explored the role of information and communication technologies (ICT) in economic growth. We discussed the principles of ICT, including digitalization and its impact on information processing and communication. Furthermore, we examined the contributions of ICT to economic growth in terms of productivity enhancement, innovation facilitation, and business transformation. We also highlighted the challenges and opportunities associated with ICT adoption, such as infrastructure requirements, the digital divide, cybersecurity concerns, and skills development. Lastly, we analyzed case studies of countries that have successfully leveraged ICT for economic development, showcasing the importance of supportive policies, infrastructure development, and digital literacy. ICT continues to shape modern economies, and understanding its role is essential for policymakers, businesses, and individuals alike.

Understanding the Importance of Biotechnology and Pharmaceuticals in Healthcare and Agriculture

Biotechnology and pharmaceuticals have revolutionized the fields of healthcare and agriculture, playing a crucial role in improving the well-being of individuals and addressing food security challenges. Biotechnology, as a multidisciplinary field, involves the use of living organisms, cells, or their components to create or modify products, processes, or systems. In healthcare, biotechnology has led to the development of innovative drugs, diagnostic tools, and therapies, while in agriculture, it has enhanced crop productivity, pest control, and genetic engineering of plants.

Biotechnology in Healthcare

Biotechnology has transformed the healthcare industry by introducing new diagnostic methods, personalized medicine, and advanced therapies. One of the most notable contributions of biotechnology is the development of recombinant DNA technology, which involves the manipulation and modification of DNA molecules. This technology has enabled the production of therapeutic proteins, such as insulin, growth factors, and antibodies, which are used to treat a wide range of diseases, including diabetes, cancer, and autoimmune disorders.

Moreover, biotechnology has significantly advanced the field of genetic engineering, allowing the modification of genes associated with diseases. Gene therapy, a cutting-edge medical procedure, involves the introduction of healthy genes into a patient's cells to correct genetic abnormalities. This approach holds immense promise for the treatment of genetic disorders and inherited diseases.

In addition to therapeutic applications, biotechnology has revolutionized diagnostics. Today, molecular diagnostics using techniques like polymerase chain reaction (PCR) and gene sequencing enable the early and accurate detection of diseases, including infectious diseases, genetic disorders, and cancer. These molecular techniques have greatly improved disease management and patient outcomes.

Biotechnology in Agriculture

Biotechnology has also played a critical role in enhancing agricultural productivity, improving crop quality, and addressing global food security challenges. Through genetic engineering, scientists can modify crop plants to exhibit desired traits, such as resistance to pests, diseases, and environmental stress. This has led to the development of genetically modified (GM) crops, which have higher yields, reduced dependence on chemical pesticides, and increased nutritional value.

For example, some GM crops are engineered to produce their own insecticide, reducing the need for extensive pesticide use. Others are modified to be resistant to herbicides, allowing for more efficient weed control. These advancements have not only increased crop yields but also minimized the environmental impact associated with traditional farming practices.

Biotechnology has also facilitated the development of improved agricultural practices. For instance, molecular markers and genetic mapping techniques allow for the identification and selection of desired plant traits, leading to the development of high-quality crops with enhanced nutritional content, longer shelf life, and better taste.

Furthermore, biotechnology has contributed to the development of vaccines for livestock, preventing the spread of infectious diseases and increasing animal health and productivity. This has a direct impact on the agricultural industry, ensuring sustainable farming practices and safeguarding animal welfare.

Challenges and Ethics in Biotechnology

While biotechnology offers immense potential in healthcare and agriculture, it also presents complex challenges and ethical considerations. One of the primary concerns is the safety of genetically modified organisms (GMOs) and their potential impact on human health and the environment. Robust regulatory frameworks and testing procedures are necessary to ensure the responsible development and commercialization of GMOs.

Additionally, the accessibility and affordability of biotechnology products and services pose challenges, particularly in developing countries. Ensuring equitable access to these technologies and addressing concerns related to intellectual property rights remain critical for widespread adoption and sustainable development.

Ethical considerations also arise when using biotechnology in healthcare, such as genetic testing and engineering. Privacy concerns, discrimination, and the potential misuse of genetic information are important ethical dilemmas that need to be addressed through stringent guidelines and regulations.

In agriculture, the impact of GM crops on biodiversity and the potential transfer of modified genes to wild plant populations are subjects of debate. Striking a balance between harnessing the potential of biotechnology and managing any associated risks is essential for the sustainable development of agriculture.

Conclusion

Biotechnology has revolutionized healthcare and agriculture, offering innovative solutions to complex challenges. Through the development of therapeutic proteins, personalized medicine, genetically modified crops, and advanced diagnostic tools, biotechnology has improved human health outcomes and agricultural productivity. However, ethical considerations and challenges related to safety, accessibility, and intellectual property rights need to be carefully addressed to ensure the responsible and equitable use of biotechnology in healthcare and agriculture. With continued research and development, biotechnology has the potential to transform these industries further, providing sustainable solutions to global challenges.

Further Reading

- Madigan, M. T., Martinko, J. M., Bender, K. S., Buckley, D. H., & Stahl, D. A. (2014). Brock biology of microorganisms (14th ed.). Pearson.
- Committee on Agricultural Biotechnology, Health, and the Environment, National Research Council. (2000). Genetically modified pest-protected plants: Science and regulation. National Academies Press.
- National Research Council. (2004). Safety of genetically engineered foods: Approaches to assessing unintended health effects. National Academies Press.
- DaSilva, E., Carter, S. R., & Sangwan, R. S. (Eds.). (2018). Advances in botanical research: Biotechnology of plant secondary metabolism (Vol. 87). Academic Press.

Analyzing the significance of nanotechnology in various sectors

Nanotechnology, the manipulation of matter at the nanoscale, has gained significant attention in recent years due to its potential impact on various sectors.

In this section, we will explore the significance of nanotechnology in sectors such as electronics, medicine, energy, and materials science. We will discuss the principles behind nanotechnology, its applications in each sector, and the challenges and opportunities it presents.

Principles of Nanotechnology

Nanotechnology revolves around the understanding and control of matter at the nanoscale, typically defined as structures with dimensions ranging from 1 to 100 nanometers. At this scale, materials exhibit unique properties and behaviors that differ from their bulk counterparts. These properties arise from quantum mechanical effects and increased surface-to-volume ratios, making nanomaterials highly versatile and desirable for various applications.

The principles of nanotechnology involve the fabrication, characterization, and manipulation of nanoscale materials. Nanomaterials can be categorized into three types: nanoparticles, nanofilms, and nanocomposites. Nanoparticles are discrete entities of nanoscale size, nanofilms are thin films with nanoscale thickness, and nanocomposites are materials composed of two or more constituents at the nanoscale.

To fabricate nanomaterials, various techniques are employed, including top-down and bottom-up approaches. The top-down approach involves breaking down bulk materials into nanoscale structures, while the bottom-up approach involves building materials atom by atom or molecule by molecule. Characterization techniques such as scanning electron microscopy (SEM) and atomic force microscopy (AFM) are used to analyze and understand the properties of nanomaterials.

Nanotechnology in Electronics

Nanotechnology has revolutionized the field of electronics by enabling the development of smaller, faster, and more efficient devices. Nanoscale materials, such as carbon nanotubes and nanowires, have shown great potential in the fabrication of nanoelectronic components.

One key application of nanotechnology in electronics is the development of transistors. Traditional transistors are made of silicon, but nanoscale transistors made of carbon nanotubes or nanowires offer advantages such as faster switching speeds, lower power consumption, and enhanced performance. These nanoscale transistors can be integrated into nanoelectronic circuits, leading to the development of high-performance electronic devices.

Another application is the use of nanomaterials in memory devices. Nanoscale magnetic materials, known as magnetic nanoparticles, have the potential to store data at higher densities compared to traditional magnetic storage devices. These nanoparticles can also be used in spintronics, where the spin of electrons is utilized to store and manipulate data. Nanotechnology plays a crucial role in the fabrication and design of these advanced memory devices.

Nanotechnology in Medicine

Nanotechnology has immense potential in the field of medicine, offering new avenues for disease diagnosis, drug delivery, and regenerative medicine. Nanomaterials with specific properties can be engineered to interact with biological systems at the cellular and molecular level, enabling precise targeting and delivery of therapeutic agents.

One area where nanotechnology has made significant advancements is in targeted drug delivery. Nanoparticles can be functionalized with targeting ligands that specifically recognize and bind to disease cells or tissues. This allows for the selective delivery of drugs, reducing side effects and improving therapeutic efficacy. Moreover, nanocarriers can protect drugs from degradation and enhance their stability in the body.

Nanotechnology also plays a crucial role in medical imaging. Nanoparticles can be engineered to have unique optical, magnetic, or radioactive properties, making them ideal contrast agents for imaging techniques such as magnetic resonance imaging (MRI), computed tomography (CT), and fluorescence imaging. These nanoparticles enable earlier detection and more accurate diagnosis of diseases.

In regenerative medicine, nanotechnology offers the potential for tissue engineering and organ transplantation. By designing nanoscale scaffolds and incorporating growth factors and cells, it is possible to fabricate artificial tissues and organs that mimic their natural counterparts. Nanotechnology also enables controlled release of growth factors, promoting tissue regeneration and repair.

Nanotechnology in Energy

Nanotechnology has the potential to revolutionize the energy sector by enabling more efficient solar cells, batteries, and fuel cells. Nanoscale materials offer unique properties that enhance energy conversion, storage, and transmission.

In the field of solar energy, nanotechnology has led to the development of more efficient photovoltaic devices. Nanoscale materials, such as quantum dots and nanowires, can be used to enhance light absorption and charge separation, improving the efficiency of solar cells. Moreover, nanomaterials can be incorporated into flexible and transparent solar cells, expanding the possibilities for their integration into various applications.

Nanotechnology also plays a crucial role in energy storage devices such as batteries. Nanoscale materials, such as nanowires and nanocomposites, can enhance the storage capacity and cycling stability of batteries. Additionally, nanomaterials can improve the performance of fuel cells by increasing the surface area and enhancing the catalytic activity of electrodes.

Furthermore, nanotechnology enables the development of energy-efficient lighting devices. Light-emitting diodes (LEDs) made of nanoscale materials, known as quantum dots, offer high-efficiency lighting with tunable colors. These nanoscale LEDs are more energy-efficient than traditional incandescent or fluorescent lights, contributing to energy savings and environmental sustainability.

Nanotechnology in Materials Science

Nanotechnology has transformed the field of materials science by enabling the development of advanced materials with enhanced properties and functionalities. Nanomaterials can exhibit superior strength, conductivity, thermal stability, and optical properties, making them ideal for a wide range of applications.

One application of nanotechnology in materials science is the development of lightweight and high-strength nanocomposites. By incorporating nanoscale reinforcements, such as carbon nanotubes or graphene, into traditional matrices, the mechanical properties of materials can be significantly improved. These nanocomposites find applications in aerospace, automotive, and construction industries.

Nanotechnology also enables the development of materials with specific surface properties. Nanoscale surface coatings, known as thin films, can be engineered to provide functionalities such as anti-reflectivity, self-cleaning, or antibacterial properties. These coatings find applications in optics, electronics, and biomedical devices.

Furthermore, nanotechnology has led to the development of innovative materials for environmental remediation. Nanoscale adsorbents, such as activated carbon nanoparticles, can remove pollutants and contaminants from air and water. Nanomaterials can also be used in sensors for the detection and monitoring of environmental pollutants, contributing to improved environmental protection and sustainability.

Challenges and Opportunities

While nanotechnology offers immense potential in various sectors, there are several challenges that need to be addressed. One major challenge is the safety and toxicity of nanomaterials. The unique properties of nanoscale materials can also pose risks to human health and the environment. It is essential to understand and mitigate these risks through rigorous safety assessments and regulations.

Another challenge is the scalability and cost-effectiveness of nanotechnology. Many nanofabrication techniques are still expensive and time-consuming, hindering large-scale production and commercialization. Efforts are underway to develop cost-effective and scalable nanomanufacturing methods to enable widespread adoption of nanotechnology.

Despite these challenges, nanotechnology presents numerous opportunities for scientific advancements and technological innovations. It opens up new possibilities for solving complex global challenges, such as clean energy production, disease treatment, and environmental sustainability. Through interdisciplinary collaborations and continued research, nanotechnology can transform various sectors and pave the way for a brighter and more sustainable future.

Conclusion

Nanotechnology holds immense significance in various sectors, including electronics, medicine, energy, and materials science. The ability to manipulate matter at the nanoscale offers unique possibilities for technological advancements and scientific breakthroughs. From smaller and faster electronic devices to targeted drug delivery, from efficient energy conversion to advanced materials with enhanced properties, nanotechnology has the potential to revolutionize multiple industries. However, addressing challenges such as safety and scalability will be crucial to realize the full potential of nanotechnology. With interdisciplinary collaborations and continuous research efforts, nanotechnology can contribute to a better and more sustainable future.

Discussing the potential of renewable energy technologies for sustainable development

Renewable energy technologies offer immense potential for sustainable development, providing clean and abundant energy sources while reducing the reliance on fossil fuels. In this section, we will explore the various renewable energy technologies and their contributions to sustainable development.

Importance of renewable energy for sustainable development

Sustainable development aims to meet the needs of the present generation without compromising the ability of future generations to meet their own needs. One of the key pillars of sustainable development is ensuring environmental sustainability. This requires a transition towards clean and renewable sources of energy to minimize the negative environmental impacts associated with fossil fuel use, such as greenhouse gas emissions and air pollution.

Renewable energy technologies offer several advantages for sustainable development:

- **Reduced greenhouse gas emissions:** Unlike fossil fuels, renewable energy sources such as solar, wind, and hydropower do not release carbon dioxide and other greenhouse gases during operation. By replacing fossil fuels with renewables, we can significantly reduce greenhouse gas emissions, mitigating climate change and promoting a greener future.

- **Energy security and independence:** Renewable energy sources are often domestically available, reducing dependency on imported fossil fuels. This enhances energy security and independence, ensuring a stable and reliable energy supply.

- **Job creation opportunities:** The renewable energy sector has the potential to create a significant number of jobs, ranging from manufacturing and installation to operation and maintenance. These job opportunities contribute to economic development and enhance social well-being.

- **Improved air quality:** Traditional energy sources like coal and oil contribute to air pollution, leading to respiratory diseases and other health hazards. Renewable energy technologies, on the other hand, produce clean energy without harmful emissions, resulting in improved air quality and better public health outcomes.

- **Diversification of energy sources:** Relying solely on fossil fuels poses risks to energy security and price volatility. By diversifying our energy sources and integrating renewables into the energy mix, we can enhance energy resilience and stability.

Considering these benefits, governments, organizations, and individuals worldwide are increasingly recognizing the potential of renewable energy technologies and taking initiatives to promote their deployment.

Major renewable energy technologies

Renewable energy technologies encompass a variety of sources, each with its own set of advantages and limitations. Let's explore some of the major renewable energy technologies contributing to sustainable development:

1. **Solar energy**: Solar energy harnesses the sun's radiation and converts it into usable electricity or heat. Photovoltaic (PV) technology utilizes solar panels to directly convert sunlight into electricity, while concentrated solar power (CSP) systems use mirrors or lenses to focus sunlight on a receiver to generate heat, which is then used to produce electricity. Solar energy is abundant, widely available, and increasingly cost-effective, with the potential to power homes, businesses, and even entire communities.

2. **Wind energy**: Wind turbines capture kinetic energy from the wind and convert it into electricity. As wind passes through the turbine blades, the rotor spins a generator, producing power. With advancements in turbine technology, wind energy has become a cost-effective and scalable solution for electricity generation. Offshore wind farms, located in coastal areas, offer even greater potential, harnessing the strong and consistent winds at sea.

3. **Hydropower**: Hydropower utilizes the energy of flowing water to generate electricity. It is one of the oldest and most widely used renewable energy sources, with hydroelectric power plants harnessing the power of rivers, dams, and even tidal movements. Hydropower offers a reliable and dispatchable source of electricity, and with careful planning, it can minimize environmental impacts on fish migration and aquatic ecosystems.

4. **Bioenergy**: Bioenergy refers to the energy derived from biomass, which includes organic matter such as wood, agricultural residues, and dedicated energy crops. Biomass can be burned directly to produce heat or electricity, or it can be converted into biofuels such as ethanol and biodiesel. Bioenergy provides a versatile and flexible renewable energy source, while also offering potential solutions for waste management and agricultural byproduct utilization.

5. **Geothermal energy**: Geothermal energy utilizes the heat stored within the Earth's crust, typically found in areas with volcanic activity or geothermal reservoirs. Geothermal power plants extract this heat and convert it into electricity. Geothermal energy is a reliable and stable source of power, with

the potential for direct use in heating and cooling applications. However, its deployment is limited to specific geological regions.

Challenges and considerations

While renewable energy technologies offer great promise, their widespread adoption faces certain challenges and considerations:

- **Intermittency and storage:** Unlike fossil fuel power plants, renewable energy sources are often intermittent, dependent on weather conditions or natural cycles. This intermittency poses challenges for grid integration and requires the development of energy storage systems to ensure a stable and reliable electricity supply.

- **Infrastructure and grid integration:** Transitioning to renewable energy requires the development of new infrastructure and the adaptation of existing grids to accommodate the fluctuating nature of renewables. Interconnecting renewable energy sources with a smart grid infrastructure enables efficient power management and grid stability while optimizing energy flows.

- **Cost considerations:** The initial investment costs associated with renewable energy technologies can be higher than conventional fossil fuel-based technologies. However, the costs of renewable energy have been steadily decreasing, making it increasingly competitive. An economic analysis should consider the long-term benefits and externalities associated with renewable energy deployment.

- **Environmental impacts:** While renewable energy sources are generally considered environmentally friendly, certain technologies can still have localized environmental impacts. For example, hydropower projects can affect aquatic ecosystems, and wind farms may have implications for bird migration patterns. Proper environmental assessments and mitigation measures are necessary to ensure sustainable deployment.

Case study: The potential of solar energy in sustainability

Solar energy represents a significant opportunity for sustainable development, with vast potential for clean and abundant electricity generation. Let's take a closer look at the case of solar energy and its impact on sustainability.

Solar energy has become increasingly affordable and accessible in recent years, thanks to advancements in photovoltaic technology, economies of scale, and supportive policies. Rooftop solar installations have gained popularity among homeowners, enabling them to generate their own electricity and reduce their carbon footprint. Moreover, large-scale solar farms can provide significant contributions to the grid, displacing fossil fuel-based generation and reducing greenhouse gas emissions.

In addition to electricity generation, solar energy can also be utilized for heating applications. Solar water heaters harness the sun's energy to heat water for domestic or commercial use, reducing the need for traditional water heating methods that rely on fossil fuels. Solar thermal energy can also be utilized in industrial processes and space heating.

Solar energy adoption brings numerous benefits to sustainability:

- **Environmental benefits:** Solar power generation produces zero greenhouse gas emissions and contributes to cleaner air quality. By displacing fossil fuel-based generation, solar energy helps mitigate the impacts of climate change and reduces local air pollution.

- **Economic opportunities:** The solar industry offers job creation opportunities across the entire value chain, from manufacturing and installation to maintenance and operation. Embracing solar energy can boost local economies and enhance energy security by reducing dependence on imported energy sources.

- **Empowering communities:** Solar energy empowers communities by enabling energy independence and resilience. Off-grid solar systems offer electricity access to remote and underserved areas, improving living conditions and supporting economic development.

- **Educational benefits:** The adoption of solar energy provides an opportunity for educational programs and awareness campaigns on renewable energy and sustainability. This helps promote a culture of environmental responsibility and encourages future generations to pursue careers in renewable energy and related fields.

Conclusion

Renewable energy technologies play a vital role in sustainable development, offering clean, abundant, and accessible sources of energy. Solar, wind, hydropower, bioenergy, and geothermal energy each contribute to the transition towards a low-carbon and sustainable future. While challenges such as intermittency, infrastructure requirements, and cost considerations exist, the benefits of renewable energy deployment outweigh the obstacles. As we continue to advance renewable energy technologies and embrace their potential, we can accelerate the path towards a greener and more sustainable world.

Key Takeaways:

- Renewable energy technologies are crucial for sustainable development, reducing greenhouse gas emissions, promoting energy security, creating jobs, and improving air quality.

- Major renewable energy technologies include solar, wind, hydropower, bioenergy, and geothermal energy.

- Challenges and considerations for renewable energy deployment include intermittency, infrastructure requirements, cost considerations, and environmental impacts.

- Solar energy holds significant potential for sustainability, offering clean electricity generation, heating applications, environmental benefits, economic opportunities, and community empowerment.

Now let's test our understanding with a couple of exercises:

Exercise 1: Compare and contrast the advantages and limitations of solar and wind energy technologies in the context of sustainable development.

Exercise 2: Discuss the potential environmental impacts associated with the large-scale deployment of hydropower projects and propose measures to mitigate these impacts.

Investigating the Impact of Advanced Manufacturing Technologies on Industrial Production

In this section, we will explore the impact of advanced manufacturing technologies on industrial production. Advanced manufacturing refers to the use of innovative technologies and processes to improve efficiency, productivity, and quality in manufacturing operations. These technologies encompass a range of tools and techniques such as automation, robotics, artificial intelligence, additive manufacturing, and data analytics. They have the potential to revolutionize industrial production and transform traditional manufacturing practices.

Importance of Advanced Manufacturing Technologies

Advanced manufacturing technologies play a crucial role in driving economic growth, competitiveness, and sustainability. They offer numerous advantages, including:

- Increased productivity: Advanced manufacturing technologies enable faster and more efficient production processes. Automation and robotics can perform repetitive tasks with precision and speed, resulting in higher output and reduced cycle times.

- Improved product quality: By leveraging advanced sensors and control systems, manufacturers can monitor and control the production process more effectively. This leads to better quality control, fewer defects, and increased customer satisfaction.

- Cost reduction: Advanced manufacturing technologies can help companies reduce operational costs through increased efficiency. Automation reduces the need for manual labor, decreasing labor costs and minimizing the risk of human error.

- Customization and flexibility: Advanced technologies allow for greater customization and flexibility in manufacturing operations. Additive manufacturing, for example, enables the production of complex and customized products without the need for expensive tooling.

- Sustainable manufacturing: With a focus on energy efficiency and waste reduction, advanced manufacturing technologies contribute to sustainable development. They help minimize the environmental impact of manufacturing processes by optimizing energy consumption and promoting the use of eco-friendly materials.

Automation and Robotics

Automation and robotics are key components of advanced manufacturing technologies. Automation involves the use of machines and systems to perform tasks autonomously, while robotics refers to the use of robotic systems to perform complex operations. These technologies have transformed manufacturing processes by enhancing efficiency, precision, and safety.

One example of automation in manufacturing is the use of robotic arms in assembly lines. These robotic arms can perform repetitive tasks with high accuracy and speed, replacing manual labor in tedious and physically demanding operations. By automating these tasks, manufacturers can achieve higher production rates and lower costs.

Furthermore, robotics has enabled the development of collaborative robots, also known as cobots. These robots are designed to work alongside human operators, assisting them in tasks that require strength, precision, or endurance. Cobots can enhance productivity and safety in manufacturing environments by taking over physically demanding or hazardous tasks, while human operators can focus on higher-level decision-making and problem-solving.

Additive Manufacturing

Additive manufacturing, also known as 3D printing, is another important advanced manufacturing technology. It involves the layer-by-layer deposition of materials to create three-dimensional objects from digital designs. Additive manufacturing has revolutionized the production process by enabling the direct fabrication of complex and customized parts.

One of the key advantages of additive manufacturing is its ability to reduce material waste. Unlike traditional subtractive manufacturing techniques, such as milling or turning, additive manufacturing only uses the necessary amount of material to create the desired object. This not only reduces waste but also allows for greater design freedom and the production of lighter, more efficient components.

Additive manufacturing also offers advantages in terms of product customization and rapid prototyping. Design modifications can be easily incorporated into the digital model, without the need for costly tooling changes. This flexibility allows manufacturers to respond quickly to customer demands and market changes, accelerating product development cycles.

Data Analytics and Artificial Intelligence

Data analytics and artificial intelligence (AI) are increasingly being applied in advanced manufacturing to optimize processes, improve quality control, and enable predictive maintenance. With the proliferation of sensors and internet connectivity, manufacturers can now collect and analyze large volumes of data in real-time.

By leveraging AI algorithms, manufacturers can extract valuable insights from this data to optimize production processes. For example, machine learning algorithms can analyze historical production data to identify patterns and anomalies, allowing for predictive maintenance and reducing machine downtime.

Furthermore, AI-powered quality control systems can automatically detect defects or deviations from specifications, ensuring consistent product quality. These systems use computer vision and pattern recognition algorithms to analyze images or sensor data, enabling real-time quality monitoring and defect identification.

Challenges and Opportunities

While advanced manufacturing technologies offer significant benefits, their implementation poses challenges and requires careful consideration. Some of the key challenges include:

- Cost of implementation: Advanced manufacturing technologies often require significant upfront investments in equipment, training, and infrastructure. Small and medium-sized enterprises (SMEs) may face difficulties in adopting these technologies due to cost constraints.

- Workforce disruption: The adoption of advanced manufacturing technologies may lead to workforce displacement or changes in job requirements. Workers may require new skills to operate and maintain these technologies, and companies need to invest in upskilling and retraining programs to ensure a smooth transition.

- Security and privacy concerns: With the increase in connectivity and data sharing, advanced manufacturing technologies raise concerns about data security and intellectual property protection. Companies need robust cybersecurity measures to protect sensitive data and ensure the integrity of their manufacturing processes.

Despite these challenges, advanced manufacturing technologies present significant opportunities for industrial production. These technologies can drive innovation, improve productivity, reduce costs, and enhance sustainability. Governments, industry associations, and academic institutions play a crucial role in promoting the adoption of these technologies and supporting manufacturers in overcoming implementation challenges.

Conclusion

Advanced manufacturing technologies have the potential to revolutionize industrial production by increasing productivity, improving product quality, and reducing costs. Automation and robotics enable efficient and precise manufacturing processes, while additive manufacturing offers flexibility, customization, and waste reduction. Data analytics and artificial intelligence optimize production processes and quality control. However, the implementation of these technologies requires careful consideration of challenges such as cost, workforce disruption, and cybersecurity concerns. By addressing these challenges and embracing the opportunities, companies can leverage advanced manufacturing technologies to gain a competitive edge and drive sustainable economic growth.

Exploring the role of artificial intelligence and machine learning in economic transformation

Artificial intelligence (AI) and machine learning (ML) have emerged as powerful technologies that are revolutionizing various sectors, including economics. In this section, we will explore the role of AI and ML in economic transformation, their applications, challenges, and potential implications for the future.

Understanding Artificial Intelligence and Machine Learning

Artificial intelligence refers to the development of computer systems that can perform tasks that typically require human intelligence. These systems are capable of learning from data, recognizing patterns, making decisions, and solving complex problems. Machine learning, a subset of AI, focuses on designing algorithms and models that enable computers to learn and improve their performance automatically without being explicitly programmed.

There are three main types of machine learning: supervised learning, unsupervised learning, and reinforcement learning. In supervised learning, models learn from labeled data to make predictions or classifications. Unsupervised learning involves analyzing unlabeled data to discover hidden patterns or structures. Reinforcement learning is based on an agent interacting with an environment, learning from feedback in the form of rewards or punishments to make optimal decisions.

Applications of Artificial Intelligence and Machine Learning

AI and ML have a wide range of applications in the economic sphere, transforming various aspects of businesses and industries. Some key applications include:

- **Automation and optimization:** AI and ML technologies enable businesses to automate routine tasks, improve operational efficiency, and optimize decision-making processes. For example, AI-powered chatbots can handle customer inquiries, while ML algorithms can optimize supply chain management operations.

- **Predictive analytics:** By analyzing large datasets, AI and ML can provide valuable insights and predictions, helping businesses make informed decisions. For instance, ML algorithms can predict customer behavior, market trends, and demand patterns, enabling businesses to optimize pricing strategies and product offerings.

- **Fraud detection and cybersecurity:** AI and ML play a crucial role in detecting and preventing fraud in financial transactions and online activities. These technologies can analyze patterns, detect anomalies, and identify potential security threats in real time, enhancing security measures and protecting businesses and consumers.

- **Personalized marketing and recommendation systems:** AI-powered recommendation systems analyze user data, preferences, and behavior to provide personalized product recommendations and targeted marketing campaigns. This enhances customer experience, increases engagement, and improves sales conversions.

- **Financial analysis and trading:** AI and ML algorithms can analyze vast amounts of financial data, predict market trends, and optimize investment strategies. High-frequency trading systems, for example, use ML algorithms to make split-second trading decisions based on real-time market data.

Challenges and Implications

While AI and ML offer significant opportunities for economic transformation, they also present challenges and potential implications.

Ethical considerations: As AI and ML technologies become increasingly integrated into economic systems, ethical considerations become critical. Issues such as data privacy, fairness in decision-making, transparency of algorithms, and AI bias need careful attention to ensure responsible and ethical AI deployment.

Labor market disruption: The automation potential of AI and ML technologies raises concerns about job displacement. Certain job sectors may experience significant changes as routine tasks are automated. However, new job opportunities may emerge, requiring different skills and expertise. Upskilling and reskilling programs are crucial to address this challenge and ensure a smooth transition in the labor market.

Data biases and quality: AI and ML algorithms heavily rely on training data for learning and decision-making. Biases present in the training data can lead to biased outcomes and decision-making processes. Ensuring high-quality, diverse, and unbiased data is essential to mitigate these issues.

Regulatory and legal frameworks: The rapid advancement of AI and ML technologies necessitates the development of robust regulatory and legal frameworks to address concerns related to privacy, security, intellectual property, liability, and accountability. Governments and policymakers need to collaborate with industry experts to establish effective regulations and standards.

Case Study: AI in the Healthcare Sector

One notable application of AI and ML is in the healthcare sector. AI algorithms can analyze medical images, such as X-rays and MRIs, for the detection of diseases, aiding radiologists in diagnosis. ML models can predict patient outcomes and help healthcare providers make personalized treatment plans. Chatbots powered by AI can assist patients in triage and provide basic medical advice.

Implementing AI in healthcare offers several benefits, including faster and more accurate diagnoses, reduced medical errors, improved patient care, and enhanced efficiency. However, challenges such as data privacy, interoperability, and regulatory compliance need to be addressed to ensure the responsible use of AI in healthcare.

Conclusion

Artificial intelligence and machine learning have transformative potential in driving economic growth and innovation. From automation and optimization to predictive analytics and personalized marketing, AI and ML are reshaping industries and revolutionizing economic systems. However, ethical considerations, labor market disruption, data biases and quality, and regulatory frameworks need to be carefully addressed to harness the full potential of these technologies. By embracing responsible AI deployment, societies can benefit from the economic transformation brought about by AI and ML.

Assessing the Potential of Blockchain Technology for Financial Inclusion and Transparency

Blockchain technology has gained significant attention in recent years for its potential to revolutionize various industries, including finance. This section explores the potential of blockchain technology in promoting financial inclusion

KEY TECHNOLOGIES FOR ECONOMIC DEVELOPMENT

and transparency. We will discuss the underlying principles of blockchain, its application in financial services, and the challenges and opportunities it presents.

Understanding Blockchain Technology

Blockchain is a decentralized and distributed ledger technology that allows for secure and transparent record-keeping. It enables the creation of a digital ledger of transactions that is shared across a network of computers, also known as nodes. These nodes work together to validate and store transactions in a sequential and immutable manner.

The key features of blockchain technology include:

- **Decentralization:** Unlike traditional centralized systems, blockchain operates on a decentralized network, removing the need for a central authority to manage and verify transactions. This distributed nature of blockchain ensures that no single entity has control over the entire system.

- **Transparency:** All transactions recorded on the blockchain are visible to every participant in the network. This transparency enhances trust and allows for real-time auditing of transactions.

- **Security:** Blockchain utilizes sophisticated cryptographic techniques to ensure the integrity and security of data. The use of hashing and digital signatures makes it extremely difficult for malicious actors to tamper with transaction records.

- **Immutability:** Once a transaction is recorded on the blockchain, it becomes virtually impossible to alter or delete. The immutability of blockchain data guarantees the integrity of the historical transaction record.

Application of Blockchain in Financial Services

Blockchain technology has the potential to address many of the challenges faced in traditional financial systems, particularly in the areas of financial inclusion and transparency. Here, we explore several key applications of blockchain in the financial services sector:

1. **Payment and Remittances:** Blockchain-based cryptocurrencies, such as Bitcoin, provide an alternative to traditional banking systems, enabling easy and efficient cross-border transactions. This can greatly benefit individuals who do not have access to traditional banking services.

2. **Smart Contracts:** Blockchain enables the execution and enforcement of smart contracts, which are self-executing agreements with the terms of the contract directly written into code. Smart contracts eliminate the need for intermediaries and can automate complex financial transactions such as loans, insurance claims, and supply chain financing.

3. **Identity Management:** Blockchain technology can facilitate the secure and decentralized management of identities. By giving individuals control over their personal data, blockchain-based identity systems can enable financial institutions to verify customer identities more efficiently, reducing the cost and time required for customer onboarding.

4. **Removal of Intermediaries:** Blockchain can eliminate intermediaries, such as clearinghouses and custodians, by providing a transparent and tamper-proof ledger. This has the potential to reduce transaction costs and streamline financial processes.

5. **Supply Chain Finance:** Blockchain can enhance transparency and trust in supply chains by securely recording transactions, verifying the authenticity of products, and tracing the movement of goods. This enables improved access to financing for small-scale suppliers who may otherwise struggle to meet traditional lending criteria.

6. **Microfinance:** Blockchain-based platforms can enable peer-to-peer lending, allowing individuals to provide loans directly to borrowers without the need for traditional financial institutions. This can expand access to affordable credit for underserved populations.

Challenges and Opportunities

While blockchain technology holds promise for financial inclusion and transparency, several challenges must be overcome for its widespread adoption. These challenges include:

- **Scalability:** The current limitations of blockchain technology, such as the time required for transaction verification and the energy consumption of consensus mechanisms, pose scalability challenges. Efforts are underway to develop solutions, such as layer-2 protocols and more energy-efficient consensus algorithms, to address these scalability issues.

- **Regulatory Frameworks:** The rapid growth of blockchain technology has outpaced the development of regulatory frameworks. The lack of clarity and consistency in regulations creates uncertainty for businesses and hampers the adoption of blockchain-based solutions.

- **Privacy and Security:** While blockchain ensures the security and integrity of transactions, it poses challenges in terms of data privacy. Striking a balance between transparency and privacy is crucial for widespread adoption.

- **Interoperability:** The lack of interoperability between different blockchain platforms and systems hinders collaboration and limits the potential benefits of blockchain technology. Efforts are being made to develop interoperability standards to enable seamless integration.

Despite these challenges, blockchain technology presents significant opportunities for financial inclusion and transparency. By leveraging its decentralized nature and cryptographic security, blockchain has the potential to empower individuals in underserved communities, provide affordable financial services, and enhance transparency in financial transactions.

Real-World Examples

Several real-world initiatives demonstrate the potential of blockchain technology for financial inclusion and transparency:

- **BitPesa:** BitPesa is a blockchain-based payments platform that enables individuals and businesses in Africa to make low-cost cross-border payments. By leveraging blockchain technology, BitPesa bypasses traditional banking systems, providing faster and cheaper remittance services to underserved populations.

- **Everledger:** Everledger is a blockchain-based platform that uses distributed ledger technology to verify the authenticity and provenance of diamonds and other high-value assets. By creating a digital record of each asset's history, Everledger promotes transparency and trust in supply chains, reducing the risk of fraud and supporting ethical sourcing.

- **Kiva:** Kiva is a blockchain-enabled microfinance platform that connects lenders with borrowers worldwide. By leveraging blockchain technology, Kiva provides transparent and secure lending opportunities for individuals in developing countries, facilitating financial inclusion and reducing poverty.

These examples demonstrate the potential of blockchain technology in promoting financial inclusion and transparency, providing a glimpse into the transformative power of decentralized and transparent systems.

Conclusion

Blockchain technology has the potential to revolutionize financial services, particularly in the areas of financial inclusion and transparency. By leveraging its decentralized nature, transparency, and security features, blockchain can enable individuals and businesses to access affordable financial services, streamline processes, and enhance trust in transactions. However, challenges, such as scalability, regulatory frameworks, privacy, and interoperability, must be addressed to realize the full potential of blockchain technology. Through continued research and innovation, blockchain has the power to foster more inclusive and transparent financial systems, ultimately contributing to sustainable economic growth and development.

Understanding the importance of cybersecurity technologies in the digital age

In today's digital age, where technology plays a vital role in various aspects of our lives, cybersecurity has become a critical concern. With the ever-increasing dependence on digital systems and networks, protecting sensitive information from unauthorized access, data breaches, and cyber threats is of utmost importance. This section aims to shed light on the importance of cybersecurity technologies in safeguarding our digital world.

The significance of cybersecurity for individuals and organizations

In an interconnected world, individuals and organizations are constantly exposed to cybersecurity risks. Cybercriminals exploit vulnerabilities in computer systems and networks to gain unauthorized access, steal valuable data, engage in identity theft, launch cyber attacks, and disrupt critical infrastructure. The consequences of such cyber threats can be devastating, ranging from financial losses and reputational damage to compromised personal safety and national security.

Cybersecurity technologies serve as a defense mechanism against these threats, providing the necessary tools and techniques to protect digital assets, ensure data confidentiality, integrity, and availability, and safeguard the privacy of individuals and organizations. By implementing robust cybersecurity measures, individuals and organizations can mitigate the risks associated with cyber attacks and create a secure digital environment for themselves and their stakeholders.

The principles of cybersecurity

Effective cybersecurity is based on a set of principles that guide the development and implementation of cybersecurity technologies. These principles include:

1. **Confidentiality**: Ensuring that information is accessible only to authorized individuals or entities.

2. **Integrity**: Maintaining the accuracy and consistency of information throughout its lifecycle.

3. **Availability**: Ensuring that information and the systems that store and process it are accessible when needed.

4. **Authentication**: Verifying the identity of users and systems to prevent unauthorized access.

5. **Authorization**: Granting appropriate privileges and permissions to users based on their roles and responsibilities.

6. **Non-repudiation**: Providing evidence that a particular action or transaction was performed, preventing individuals from denying their involvement.

7. **Accountability**: Holding individuals or entities responsible for their actions and ensuring traceability.

Adhering to these principles is critical in designing secure systems and networks that can withstand cyber threats and protect sensitive information.

Cybersecurity technologies and their applications

Cybersecurity technologies encompass a wide range of tools, technologies, and practices designed to protect information systems and networks. Here are some key cybersecurity technologies and their applications:

1. **Firewalls**: Firewalls act as a barrier between internal networks and external networks, monitoring and controlling incoming and outgoing network traffic based on predefined security rules. They help prevent unauthorized access and protect against network-based attacks.

2. **Intrusion Detection Systems (IDS)** and **Intrusion Prevention Systems (IPS)**: IDS and IPS are security solutions that detect and prevent malicious activities on a network. IDS monitors network traffic and alerts administrators when suspicious activity is detected, while IPS takes immediate action to block and prevent the unauthorized activity from occurring.

3. **Encryption:** Encryption is the process of converting plain text information into an unreadable format, known as ciphertext, using mathematical algorithms. It ensures the confidentiality and integrity of data by making it unreadable to unauthorized parties. Encryption is widely used in securing sensitive communication, such as online transactions and email exchanges.

4. **Antivirus and Antimalware Software:** Antivirus and antimalware software detect, prevent, and remove malicious software, such as viruses, worms, and spyware, from computer systems. These software solutions constantly monitor for known patterns and behaviors of malware to protect against infections.

5. **Access Control Systems:** Access control systems allow organizations to control and restrict access to their systems and networks based on user roles, privileges, and authentication mechanisms. This helps prevent unauthorized users from gaining access to sensitive information.

6. **Security Information and Event Management (SIEM):** SIEM solutions provide real-time monitoring, correlation, and analysis of security events and incidents, helping security professionals detect and respond to potential threats. SIEM technologies collect logs and events from various sources, allowing for centralized management and analysis of security-related data.

Ethical considerations in cybersecurity

While cybersecurity technologies play a critical role in protecting digital assets, it is essential to consider the ethical implications associated with their usage. Here are some key ethical considerations in cybersecurity:

1. **Privacy:** As cybersecurity technologies collect and analyze vast amounts of data, it is crucial to balance the need for security with the privacy rights of individuals. Organizations must handle personal information responsibly and ensure compliance with privacy regulations.

2. **Transparency**: It is important for organizations to be transparent about their cybersecurity practices, including the types of data collected, how it is used, and who has access to it. Open and transparent communication builds trust and accountability.

3. **Accountability**: Organizations should take responsibility for their cybersecurity practices and be accountable for any breaches or incidents. This includes timely disclosure of breaches, appropriate remediation measures, and compensating affected individuals or entities.

4. **Proportionality**: Cybersecurity measures should be proportionate to the risks and threats faced by organizations. Excessive surveillance or restrictive measures that infringe on individual freedoms without sufficient justification should be avoided.

5. **Collaboration**: Cybersecurity is a collective effort, and collaboration among organizations, governments, and individuals is crucial. Sharing threat intelligence, best practices, and lessons learned fosters a stronger cybersecurity ecosystem.

Real-world example: The Equifax data breach

The Equifax data breach in 2017 serves as a compelling example of the importance of cybersecurity technologies. Equifax, one of the largest credit reporting agencies, experienced a massive data breach that exposed the sensitive information of approximately 147 million consumers. The breach resulted from a failure to patch a known software vulnerability, allowing cybercriminals to gain unauthorized access to the company's systems.

This incident highlights the pressing need for organizations to prioritize cybersecurity and implement robust technologies and practices to protect sensitive information. It also underscores the significance of cybersecurity in maintaining trust and confidence in our digital ecosystem.

Conclusion

As technology continues to advance and our reliance on digital systems grows, the importance of cybersecurity technologies cannot be understated. These technologies play a crucial role in safeguarding our digital assets, protecting sensitive information, and mitigating the risks associated with cyber threats. By understanding the significance of cybersecurity and adopting appropriate technologies and practices, individuals and organizations can navigate the digital age with confidence and peace of mind.

Key Takeaways

1. Cybersecurity technologies are essential for protecting individuals and organizations from cyber threats and ensuring the confidentiality, integrity, and availability of digital assets.

2. The principles of confidentiality, integrity, availability, authentication, authorization, non-repudiation, and accountability form the foundation of effective cybersecurity.

3. Key cybersecurity technologies include firewalls, intrusion detection and prevention systems, encryption, antivirus and antimalware software, access control systems, and security information and event management systems.

4. Ethical considerations in cybersecurity include privacy, transparency, accountability, proportionality, and collaboration.

5. Real-world incidents, such as the Equifax data breach, highlight the critical importance of cybersecurity technologies in protecting sensitive information.

Discussion Questions

1. What are the potential consequences of a cybersecurity breach for individuals and organizations? Provide examples.

2. Explain the concept of encryption and its significance in cybersecurity.

3. Discuss the ethical concerns associated with the use of cybersecurity technologies.

Analyzing the role of space technology and satellite communications in economic development

Space technology and satellite communications play a vital role in economic development. They have revolutionized various sectors and have become essential tools for businesses, governments, and individuals worldwide. In this section, we will explore the significance of space technology and satellite communications in fostering economic growth and how they have transformed industries, improved connectivity, and contributed to sustainable development.

Understanding Space Technology

Space technology refers to the use of various systems, devices, and instruments to explore and utilize space for practical purposes. It encompasses a wide range of applications, including satellite systems, space probes, telescopes, and space stations. The development and deployment of space technology have made significant contributions to scientific research, communication, navigation, weather forecasting, and resource management.

The Role of Satellite Communications

Satellite communications involve the use of artificial satellites to transmit and receive signals for communication purposes. These satellites are positioned in geostationary or low-earth orbits and provide coverage over large areas, enabling global communication and connectivity. Satellite communications have revolutionized telecommunications, broadcasting, and internet services, significantly impacting economic development in various ways:

1. **Global Connectivity:** Satellite communications have connected people and businesses globally, overcoming geographical barriers. They have provided access to telecommunications services in remote and underserved areas, enabling communication, data transfer, and internet connectivity. This connectivity has facilitated trade, e-commerce, and collaboration across borders, contributing to economic growth.

2. **Telecommunications and Broadcasting:** Satellite communications have transformed the telecommunications and broadcasting industries. They enable the transmission of voice, data, and video signals over long distances, ensuring reliable and efficient communication. Satellite television and radio broadcasting have expanded access to information, entertainment, and educational content, fostering cultural exchange and economic opportunities.

3. **Emergency and Disaster Management:** Satellite communications play a critical role in emergency and disaster management. During emergencies, such as natural disasters or humanitarian crises, satellite communication systems provide essential communication links, enabling coordination, rescue operations, and the delivery of humanitarian aid. This capability helps mitigate the impact of disasters and supports recovery efforts, contributing to economic resilience.

4. **Navigation and Positioning:** Satellite-based navigation systems, such as the Global Positioning System (GPS), have become integral to various sectors, including transportation, logistics, and agriculture. GPS enables accurate positioning, navigation, and timing information, enhancing efficiency, safety, and productivity. Businesses can optimize their supply chains, vehicle routing, and fleet management, leading to cost reduction and improved competitiveness.

5. **Earth Observation and Resource Management:** Space technology, including satellite imagery and remote sensing, allows for monitoring and management of Earth's resources. Satellites provide valuable data on climate patterns, weather forecasting, natural resource exploration, environmental monitoring, and urban planning. This information is crucial for sustainable resource management, disaster risk reduction, and informed decision-making, benefiting economic sectors such as agriculture, forestry, energy, and mining.

Innovation and Entrepreneurship in Space Technology

The development and utilization of space technology have opened up new opportunities for innovation and entrepreneurship. As space technology becomes more accessible and affordable, we are witnessing a rapid increase in startups and private companies venturing into space-related activities. These ventures focus on developing innovative solutions using space technology, such as satellite-based applications, space tourism, and asteroid mining.

Governments worldwide are also incentivizing space entrepreneurship through policies and funding programs. This push for commercial space activities has the potential to create new industries, generate employment opportunities, and stimulate economic growth. It fosters innovation in satellite design, launch systems, data analytics, and applications, driving technological advancements and economic competitiveness.

Challenges and Ethical Considerations

While space technology and satellite communications offer significant benefits for economic development, they also pose challenges and ethical considerations. Some key challenges include:

- **High Cost:** Building and launching satellites and maintaining ground infrastructure for satellite communications can be expensive. The high cost of space technology limits access and adoption, especially for developing countries and small businesses.

- **Space Debris and Sustainability:** The increasing number of satellites and space debris in orbit poses risks of collisions and impacts on existing infrastructure. Managing space debris and ensuring sustainability in space activities are critical challenges for the long-term viability of space technology.

- **Privacy and Security:** Satellite communications involve the transmission and storage of sensitive data. Ensuring privacy and cybersecurity in satellite systems is essential to protect individuals and businesses from unauthorized access or data breaches.

- **Unequal Access and the Digital Divide:** While satellite communications provide connectivity in remote and underserved areas, unequal access to satellite-based services and technologies can contribute to a digital divide. Bridging this divide is crucial for promoting inclusive economic development.

Addressing these challenges requires collaboration between governments, international organizations, and the private sector. It entails supporting research and development, promoting technology transfer, investing in infrastructure, and formulating policies that prioritize sustainability, equity, and responsible space activities.

Case Study: Impact of Satellite Communications in Remote Healthcare

An excellent example of the role of space technology and satellite communications in economic development is remote healthcare delivery. In remote and underserved areas where healthcare infrastructure is limited, satellite communications enable telemedicine and telehealth services. These services connect healthcare professionals with patients through video consultations, remote diagnosis, and the transfer of medical data, ensuring access to quality healthcare.

Satellite-enabled remote healthcare helps overcome barriers such as distance, lack of infrastructure, and scarcity of medical professionals. It improves healthcare outcomes, reduces costs, and enhances the efficiency of healthcare delivery, particularly in emergencies and resource-constrained environments. This technology has significant socioeconomic impacts, enabling better healthcare access, saving lives, and contributing to economic productivity and well-being.

Conclusion

Space technology and satellite communications have become integral to economic development, fostering innovation, connectivity, and sustainable resource management. They have transformed industries, facilitated global connectivity, enhanced emergency management, improved navigation and positioning, and enabled remote services like telemedicine. However, addressing challenges such as cost, sustainability, privacy, and unequal access is crucial to ensuring the responsible and equitable use of space technology for economic growth. Governments, international organizations, and the private sector must collaborate to leverage the benefits of space technology while promoting ethical and sustainable practices in space activities.

Discussing the opportunities and challenges of emerging technologies for economic growth

Emerging technologies play a crucial role in driving economic growth and shaping the future of industries across various sectors. These technologies, characterized by their novelty and potential for disruptive impact, offer both opportunities and challenges for businesses and economies. In this section, we will explore the potential benefits and risks associated with the adoption and diffusion of emerging technologies, and discuss strategies to harness their power for sustainable economic development.

Opportunities for Economic Growth

1. Enhanced Efficiency and Productivity: One of the key opportunities presented by emerging technologies is the potential to significantly improve efficiency and productivity. For example, automation technologies and artificial intelligence (AI) systems can streamline processes, reduce manual labor, and optimize resource allocation. This can lead to cost savings, increased output, and improved competitiveness for businesses.

2. Innovation and Market Disruption: Emerging technologies often bring about disruptive innovations that can reshape industries, create new markets, and drive economic growth. For instance, breakthroughs in biotechnology and pharmaceuticals have revolutionized healthcare by enabling personalized medicine, gene editing, and advanced diagnostic techniques. Similarly, advancements in renewable energy technologies have facilitated the transition towards a sustainable and low-carbon economy.

3. Job Creation and Skill Development: While emerging technologies may automate certain tasks, they also create new job opportunities and foster the development of new skills. For instance, the implementation of Industry 4.0 technologies, such as the Internet of Things (IoT) and advanced robotics, has the potential to create jobs in fields such as data analytics, cybersecurity, and software development. Additionally, emerging technologies often require a skilled workforce, offering opportunities for upskilling and reskilling initiatives.

4. Economic Inclusion and Social Impact: Emerging technologies have the potential to bridge socioeconomic gaps and promote economic inclusion. For example, digital platforms and mobile technologies have enabled the growth of the sharing economy, providing opportunities for small businesses and individuals to access new markets. Furthermore, technologies like blockchain have the potential to enhance transparency, traceability, and accountability, thereby reducing corruption and promoting inclusive economic development.

5. Global Collaboration and Knowledge Sharing: The adoption of emerging technologies can foster global collaboration, knowledge sharing, and international partnerships. For instance, advancements in communication technologies enable real-time collaboration among researchers, innovators, and entrepreneurs from different parts of the world. This can facilitate the cross-pollination of ideas, accelerate innovation cycles, and drive economic growth through global networks.

Challenges for Economic Growth

1. Skills Gap and Workforce Transformation: The rapid pace of technological advancements creates a challenge in terms of the skills gap and workforce transformation. The adoption of emerging technologies often requires specialized knowledge and skills that are in short supply. This calls for investments in education and training programs to ensure a skilled workforce capable of leveraging the opportunities presented by these technologies.

2. Security and Privacy Concerns: The widespread adoption of emerging technologies raises concerns about data security and privacy. For example, the proliferation of IoT devices poses cybersecurity risks, as they become potential targets for hackers. Similarly, AI applications that rely on extensive data collection and analysis must address privacy concerns to build user trust and ensure responsible use of personal information.

3. Ethical and Social Implications: Emerging technologies, such as AI and automation, raise ethical questions regarding job displacement, algorithmic bias, and the concentration of power in the hands of a few. Addressing these concerns

requires the development of ethical frameworks, regulations, and policies that ensure the responsible and equitable use of emerging technologies.

4. Infrastructure and Access: The successful adoption of emerging technologies requires robust technological infrastructure, including reliable internet connectivity, energy supply, and digital platforms. However, access to infrastructure is not uniform across countries and regions, creating a digital divide that hampers inclusive growth. Bridging this gap requires investments in infrastructure development and policies that prioritize equitable access to emerging technologies.

5. Economic Disruption and Unequal Distribution: Disruptive innovations can lead to the displacement of traditional industries and jobs, which may create temporary economic disruption and increase income inequality. It is important to implement policies and mechanisms that mitigate the negative impacts of these disruptions and ensure a just transition for affected individuals and communities.

6. Financial Investment and Uncertainty: The adoption of emerging technologies often requires substantial financial investment, especially in research, development, and infrastructure. However, the rapid pace of technological change and the uncertainty associated with emerging technologies pose risks for investors. Governments and private sector stakeholders need to collaborate to provide financial support and create an enabling environment for investment in emerging technologies.

Strategies for Harnessing Emerging Technologies

1. Promoting Research and Development: Governments, academia, and the private sector should invest in research and development efforts aimed at advancing emerging technologies. This includes funding research projects, establishing innovation hubs and centers of excellence, and creating collaborative networks for knowledge sharing.

2. Developing Digital Skills: To address the skills gap, there is a need to prioritize digital skills development and provide training opportunities for individuals and businesses. This can be achieved through partnerships between educational institutions, industry associations, and government agencies, focusing on emerging technology-related skills such as data analytics, AI, and cybersecurity.

3. Encouraging Collaboration and Knowledge Sharing: Governments and international organizations should facilitate collaboration and knowledge sharing among different stakeholders, including researchers, policymakers, entrepreneurs, and industry leaders. This can be achieved through conferences, workshops, and

international partnerships that promote interdisciplinary collaboration and the exchange of best practices.

4. Establishing Regulatory Frameworks: Ethical and responsible use of emerging technologies requires the development of regulatory frameworks that address concerns related to privacy, security, fairness, and accountability. Policymakers should collaborate with experts, industry representatives, and civil society organizations to establish guidelines and regulations that ensure the safe and beneficial deployment of emerging technologies.

5. Bridging the Digital Divide: Addressing the digital divide requires investments in technological infrastructure, particularly in underserved areas. Governments can provide incentives for private sector investment, establish public-private partnerships, and develop policies that promote digital inclusion and equitable access to emerging technologies.

6. Fostering Entrepreneurship and Innovation: Governments can create favorable conditions for technology startups by providing financial support, regulatory incentives, and access to networks and resources. Incubators and accelerators can play a crucial role in nurturing technology-driven entrepreneurship, supporting startups, and connecting them with mentors and investors.

7. Ensuring Ethical and Inclusive Development: Policymakers and technologists need to prioritize the principles of ethics, inclusion, and sustainability in the development and deployment of emerging technologies. This includes addressing biases in AI algorithms, ensuring transparency and accountability, and considering social and environmental impacts in technological decision-making.

Conclusion

Emerging technologies offer immense opportunities for economic growth, innovation, and social progress. However, they also present a range of challenges that must be addressed to maximize their potential and ensure inclusive and sustainable development. By adopting the strategies outlined in this section, policymakers, businesses, and society can navigate the complex landscape of emerging technologies and harness their transformative power for the benefit of all.

Technology Adoption and Diffusion

Understanding the process of technology adoption

Exploring the factors influencing technology adoption by individuals

Technology adoption plays a crucial role in the overall process of technological transformation and economic development. In this section, we will explore the various factors that influence individuals' decisions to adopt new technologies. Understanding these factors is essential for policymakers, businesses, and other stakeholders to design effective strategies to promote technology adoption and drive economic growth.

Awareness and Perceived Benefits

One of the key factors that influence technology adoption is individuals' awareness of the technology and their perception of its benefits. People are more likely to adopt a new technology if they have sufficient knowledge about it and believe that it can improve their lives, work, or overall well-being. For example, an individual may be more inclined to adopt a new smartphone if they are aware of its features and perceive benefits such as enhanced communication, access to information, and convenience.

Perceived Ease of Use

The perceived ease of use of a technology is another critical factor in adoption. If individuals find a technology easy to understand, learn, and operate, they are more likely to adopt it. On the other hand, technologies that are perceived as complex or difficult to use may deter individuals from adopting them. User-friendly interfaces, intuitive controls, and clear instructions can significantly impact the perceived ease of use and influence adoption rates.

Costs and Affordability

The cost of technology and its affordability are significant considerations in adoption decisions. Individuals assess the financial implications of adopting a new technology, including the initial purchase cost, ongoing maintenance expenses, and potential upgrades or replacements. The affordability of technology varies according to individuals' income levels, disposable income, and the availability of financing options. Lower-cost alternatives or subsidies can make technology more accessible and increase adoption rates.

Compatibility with Existing Systems

Compatibility with existing systems or technologies is another crucial factor influencing adoption. Individuals may hesitate to adopt a new technology if it is not compatible with their current devices, software, or infrastructure. Compatibility issues can create inconvenience, additional costs, or disruptions, making individuals reluctant to adopt new technologies. Compatibility standards and interoperability can play a significant role in driving technology adoption.

Social Influences

Social influences, such as social norms, peer pressure, and recommendations from friends or family, can strongly impact technology adoption. Individuals often seek validation from their social networks before making adoption decisions. Positive experiences and recommendations from trusted individuals or influential groups can encourage technology adoption, while negative perceptions or social stigma may hinder it. Understanding social dynamics and leveraging social networks can be effective strategies for promoting technology adoption.

Perceived Risks and Uncertainty

Perceived risks and uncertainties associated with a new technology can act as significant barriers to adoption. Individuals may fear potential negative consequences, such as privacy breaches, security threats, or negative health effects. Perceived risks can be influenced by media coverage, personal experiences, and the level of trust in the technology or its providers. Clear communication about risks, effective risk mitigation strategies, and regulatory frameworks can address individuals' concerns and encourage technology adoption.

Education and Digital Literacy

Education and digital literacy are essential factors influencing technology adoption. Individuals with higher levels of education and digital skills are typically more comfortable and confident in adopting new technologies. Lack of digital literacy or access to education can hinder individuals' ability to use and adopt technology. Providing training programs, digital literacy initiatives, and affordable access to technology can bridge the digital divide and promote technology adoption.

Infrastructure and Accessibility

The availability and quality of technological infrastructure, including internet connectivity, electricity, and supporting services, significantly influence technology adoption. Individuals in regions with poor or limited infrastructure may face challenges in accessing and adopting new technologies. Investment in infrastructure development, expansion of connectivity, and accessibility initiatives can play a pivotal role in promoting technology adoption.

Cultural and Regulatory Factors

Cultural and regulatory factors also shape individuals' attitudes towards technology adoption. Cultural norms, values, and beliefs can influence perceptions of technology and its role in society. Regulatory frameworks, including laws, standards, and policies, can either facilitate or hinder adoption. Understanding cultural nuances and aligning regulatory frameworks with technological advancements can encourage technology adoption.

Incentives and Rewards

Incentives and rewards can motivate individuals to adopt new technologies. Financial incentives, such as tax credits or subsidies, can reduce the financial burden of adoption. Non-financial incentives, such as loyalty programs or recognition, can also encourage adoption. Businesses and policymakers can leverage incentives and rewards to drive technology adoption and create a positive feedback loop.

Real or Perceived Barriers

Individuals may face real or perceived barriers to technology adoption, such as limited access to technology, lack of technical support, or concerns about privacy and security. Identifying and addressing these barriers is crucial in fostering technology adoption. Strategies, including improving access, providing technical support, addressing privacy concerns, and building trust, can eliminate or mitigate these barriers and promote adoption.

In conclusion, several interrelated factors influence individuals' decisions to adopt new technologies. Awareness, perceived benefits, ease of use, costs, compatibility, social influences, risks, education, infrastructure, cultural factors, regulation, incentives, and barriers all play a significant role. Policymakers and businesses should consider these factors when designing strategies to promote technology adoption and create an environment conducive to economic growth and development.

Key Takeaways

- Individuals' decisions to adopt new technologies are influenced by factors such as awareness, perceived benefits, ease of use, costs, compatibility, social influences, risks, education, infrastructure, cultural factors, regulation, incentives, and barriers.
- Promoting technology adoption requires addressing these factors through strategies such as clear communication, user-friendly interfaces, affordability measures, compatibility standards, leveraging social networks, risk mitigation, education, infrastructure development, cultural alignment, regulatory frameworks, incentives, and barrier mitigation.
- Effective technology adoption strategies can drive economic growth and development by enabling individuals to harness the benefits of new technologies for personal, professional, and societal advancement.

Analyzing the barriers to technology adoption in organizations

Technology adoption in organizations is often influenced by various barriers that hinder the successful implementation and integration of new technologies. These barriers can arise from internal factors within an organization or external factors in the broader business environment. In this section, we will analyze some of the common barriers to technology adoption in organizations and discuss strategies for overcoming them.

Lack of technological infrastructure

One of the primary barriers to technology adoption in organizations is the lack of suitable technological infrastructure. This includes the absence of necessary hardware, software, and network infrastructure required for the new technology to operate effectively. Companies may not have the financial resources to invest in infrastructure upgrades or lack the technical expertise to implement and maintain the required technological infrastructure.

Organizations can overcome this barrier by conducting a thorough assessment of their existing technological infrastructure and identifying the gaps that need to be addressed. They can then develop a comprehensive plan to acquire the necessary hardware, software, and network infrastructure. Collaboration with technology vendors, seeking external funding sources, or leveraging cloud-based solutions can also help organizations overcome infrastructure-related barriers.

Resistance to change

Resistance to change is a significant barrier to technology adoption in organizations. Employees may feel apprehensive about learning new technologies and fear that their job roles may be negatively impacted. Organizational culture and resistance from key stakeholders can also contribute to this barrier.

To address resistance to change, organizations need to create a supportive and inclusive environment. This can be achieved through effective change management strategies, such as clear communication about the benefits of the new technology, involving employees in the decision-making process, providing comprehensive training programs, and offering incentives for embracing the change. Additionally, fostering a culture of continuous learning and innovation can help overcome resistance to change.

Lack of digital skills and knowledge

Another common barrier to technology adoption in organizations is the lack of digital skills and knowledge among employees. Many technologies require specific technical competencies, and the absence of these skills can hinder the successful implementation and utilization of new technologies.

To overcome this barrier, organizations should invest in training and development programs to enhance digital literacy and skills. This can include providing workshops, online courses, and hands-on training sessions. Collaboration with educational institutions, industry associations, or technology service providers can also help organizations bridge the digital skills gap by offering specialized training programs.

Financial constraints

Financial constraints can pose a significant barrier to technology adoption, especially for small and medium-sized organizations. Implementing new technologies often requires substantial upfront investments in hardware, software, implementation, and training costs. Limited financial resources may hinder organizations from adopting new technologies, particularly if the return on investment is not immediately apparent.

Organizations can overcome financial constraints by exploring alternative funding options, such as government grants or loans specifically designed to support technology adoption. They can also consider leasing options or partnerships with technology vendors that offer flexible payment plans. Conducting a cost-benefit analysis to demonstrate the long-term value and potential cost savings of the new technology can also facilitate securing financial resources.

Data security and privacy concerns

Data security and privacy concerns are significant barriers to technology adoption, particularly in industries that handle sensitive information. Organizations may be reluctant to adopt new technologies due to concerns about data breaches, unauthorized access, or compliance with data protection regulations.

To address data security and privacy concerns, organizations should prioritize cybersecurity measures and data protection protocols. This can include implementing encryption techniques, securing network infrastructure, training employees on data security best practices, and regularly auditing and monitoring data systems. Compliance with applicable data protection regulations, such as the General Data Protection Regulation (GDPR), can also help alleviate concerns and build trust with stakeholders.

Interoperability and integration challenges

Interoperability and integration challenges can serve as significant barriers to technology adoption, especially when organizations have existing systems and processes in place. Incompatibility between new and legacy systems, difficulties in integrating different technologies, and data migration issues can impede the successful adoption of new technologies.

To overcome interoperability and integration challenges, organizations should conduct a thorough analysis of their existing systems and assess compatibility with the new technology. Collaboration with technology vendors to develop customized integration solutions or leveraging application programming interfaces (APIs) can help streamline the integration process. Additionally, organizations should prioritize data standardization and establish clear protocols for data migration to ensure smooth transitioning.

Regulatory and compliance requirements

Regulatory and compliance requirements can present significant barriers to technology adoption, particularly in highly regulated industries such as finance, healthcare, and government. Organizations may hesitate to adopt new technologies due to the lack of clarity regarding regulatory compliance or concerns about legal implications.

To address regulatory and compliance requirements, organizations should collaborate with legal and regulatory experts to understand the implications of adopting new technologies. Ensuring transparency and accountability in data handling, implementing robust cybersecurity measures, and complying with industry-specific regulations are essential steps in overcoming this barrier. Engaging in proactive communication with regulatory bodies and seeking guidance can also help in navigating compliance challenges.

In conclusion, various barriers can hinder the successful adoption of new technologies in organizations. Lack of technological infrastructure, resistance to change, lack of digital skills, financial constraints, data security concerns, interoperability challenges, and regulatory requirements are among the common barriers. However, organizations can overcome these barriers through strategic planning, clear communication, training and development programs, financial planning, cybersecurity measures, and collaboration with relevant stakeholders. By addressing these barriers, organizations can unlock the potential benefits of technology adoption and drive innovation and growth.

Discussing the role of government policies in promoting technology adoption

Government policies play a crucial role in promoting technology adoption within a country. These policies provide a framework for businesses and individuals to embrace and incorporate new technologies, thereby driving economic growth and innovation. In this section, we will examine the various ways in which government policies can facilitate technology adoption and discuss their implications.

Understanding the importance of technology adoption

Technology adoption refers to the process through which individuals, organizations, and societies incorporate new technologies into their daily operations and practices. It enables them to enhance their efficiency, productivity, and competitiveness, thereby contributing to economic development. However, technology adoption can be a complex and challenging task, requiring significant investments and changes in organizational and individual behavior. This is where government policies play a vital role in providing incentives, support, and guidance to foster technology adoption.

Examining the factors influencing technology adoption by individuals

Individuals are key drivers in the adoption of technology. However, there are various factors that influence their decision-making process. These factors can include perceived usefulness, ease of use, cost, access, and familiarity with the technology. Government policies can address these factors through measures such as providing subsidies or tax incentives to individuals who adopt specific technologies. Additionally, public awareness campaigns and educational programs can be implemented to familiarize individuals with the benefits and usage of new technologies.

For example, in the case of renewable energy technologies, the government can introduce policies that provide financial incentives such as feed-in tariffs or tax credits to individuals who adopt solar panels or install energy-efficient appliances. These policies reduce the financial burden of adopting new technologies and encourage individuals to embrace sustainable practices.

Analyzing the barriers to technology adoption in organizations

Organizations often face barriers when it comes to incorporating new technologies into their operations. These barriers can include high implementation costs, lack of technical knowledge or expertise, resistance to change, and uncertainties regarding the return on investment. Government policies can address these barriers through various measures.

One effective approach is the provision of financial incentives and support programs for technology adoption. For instance, governments can offer grants or low-interest loans to incentivize businesses to invest in advanced manufacturing technologies or adopt digital solutions to improve their processes. Additionally, the government can establish partnerships with educational institutions and industry associations to offer training programs and technical support to organizations, enabling them to overcome technical barriers.

Discussing the role of government policies in promoting technology adoption

Government policies play a crucial role in promoting technology adoption by creating an enabling environment for businesses and individuals. By implementing supportive policies, governments can facilitate the diffusion of new technologies across various sectors. Some key policy measures include:

1. **Research and Development (R&D) funding**: Governments can allocate funding for R&D activities in universities, research institutions, and private enterprises. This funding helps to develop new technologies and innovations, which can subsequently be adopted by businesses and individuals. It encourages collaboration between academia and industry, fostering technological advancements and commercialization.

2. **Tax incentives and subsidies**: Governments can provide tax incentives and subsidies to reduce the financial burden of technology adoption. This can include tax credits for investments in research and development, allowances for the purchase of advanced technologies, or direct subsidies for the adoption of specific technologies. These incentives encourage businesses to invest in new technologies and drive their adoption.

3. **Regulatory frameworks and standards**: Governments can establish regulatory frameworks and set industry standards to ensure the compatibility, interoperability, and safety of technologies. These regulations provide clarity and confidence to businesses and individuals, fostering trust in the adoption process. For example, in the telecommunications sector, governments play a crucial role in allocating and regulating the use of frequency spectrum, which promotes the development and adoption of wireless technologies.

4. **Supportive infrastructure development**: Governments can invest in the development of necessary infrastructure to support technology adoption. This includes the establishment of broadband and telecommunications networks, data centers, and research facilities. By providing a robust

technological infrastructure, governments facilitate the adoption and utilization of new technologies by businesses and individuals.

5. **Education and skill development:** Governments can invest in education and skill development programs to ensure a competent and skilled workforce that is capable of adopting and utilizing new technologies. This includes promoting science, technology, engineering, and mathematics (STEM) education, offering vocational training programs, and supporting research and innovation in educational institutions. By enhancing the technological capabilities of the workforce, governments enable businesses and individuals to effectively adopt and harness the benefits of new technologies.

Understanding the ethical implications of technology adoption and diffusion

While technology adoption brings numerous benefits, it also raises ethical concerns that need to be addressed by government policies. Some of the key ethical implications include:

- **Privacy and data protection:** With the increasing adoption of technologies such as artificial intelligence and the internet of things, the collection and utilization of personal data have become widespread. Governments need to establish robust regulations and frameworks to protect individuals' privacy rights and ensure the responsible use of personal data.

- **Equity and access:** Technology adoption can exacerbate existing inequalities if certain segments of the population do not have equal access to new technologies. Governments need to implement policies that ensure equitable access to technology, reducing the digital divide and promoting inclusivity.

- **Job displacement and workforce transitions:** The adoption of technologies like automation and artificial intelligence can lead to job displacement for certain sectors of the workforce. Governments must develop policies and programs to support affected individuals in transitioning to new job opportunities or acquiring new skills needed for emerging technologies.

- **Security and cyber threats:** As technology adoption increases, the risks posed by cybersecurity threats and data breaches also rise. Governments need to strengthen cybersecurity measures, establish legal frameworks for addressing cybercrimes, and promote awareness and education regarding online security.

- **Environmental sustainability:** Technology adoption should align with sustainable development principles to minimize negative environmental impacts. Governments must promote the adoption of sustainable technologies and establish policies that encourage businesses and individuals to embrace environmentally friendly practices.

Discussing strategies for fostering inclusive and responsible technology adoption

In order to ensure inclusive and responsible technology adoption, governments can implement the following strategies:

- **Broad-based awareness and education:** Governments can launch public awareness campaigns to educate individuals and businesses about the benefits and potential risks of adopting new technologies. These campaigns can provide information on available support programs, guidelines for responsible technology adoption, and awareness of ethical considerations.

- **Partnerships and collaborations:** Governments should engage in partnerships and collaborations with industry players, academia, and civil society organizations to jointly develop strategies and solutions for promoting technology adoption. This promotes knowledge sharing, fosters innovation, and ensures that policies are informed by diverse perspectives.

- **Inclusive policy development:** Government policies should take into account the needs and concerns of all stakeholders, including marginalized communities and vulnerable groups. Policies should be designed to address inequalities, reduce barriers to technology adoption, and promote inclusivity.

- **Continuous monitoring and evaluation:** Governments should establish mechanisms for monitoring and evaluating the impact of technology adoption policies. This allows for the identification of any unintended consequences or areas requiring improvement, enabling policymakers to make informed decisions and adjust policies accordingly.

In conclusion, government policies play a pivotal role in promoting technology adoption by creating an enabling environment, addressing barriers, and providing incentives for individuals and organizations. By implementing supportive policies, governments can drive economic growth, foster innovation, and ensure that technology adoption is inclusive, responsible, and aligned with ethical considerations. It is through these policies that countries can effectively harness the potential of new technologies and achieve sustainable development.

Key Takeaways

- Government policies play a crucial role in promoting technology adoption by creating an enabling environment, addressing barriers, and providing incentives for individuals and organizations.

- Factors influencing technology adoption by individuals include perceived usefulness, ease of use, cost, access, and familiarity with the technology.

- Organizational barriers to technology adoption include high implementation costs, lack of technical knowledge, resistance to change, and uncertainties regarding return on investment.

- Key government policies for promoting technology adoption include R&D funding, tax incentives and subsidies, regulatory frameworks, infrastructure development, and education and skill development.

- Ethical implications of technology adoption include privacy and data protection, equity and access, job displacement, security and cyber threats, and environmental sustainability.

- Strategies for fostering inclusive and responsible technology adoption include awareness and education, partnerships and collaborations, inclusive policy development, and continuous monitoring and evaluation.

Resources

- World Bank, "Digital Adoption: How to Help Businesses Get Digital" [Online]. Available: https://www.worldbank.org/en/topic/ict/brief/digital-adoption

- European Commission, "Accelerating the Uptake of Emerging Technologies across EU's Economy" [Online]. Available: https://ec.europa.eu/transparency/regdoc/rep/1/2020/EN/COM-2020-237-F1-EN-ANNEX-1-PART-1.PDF

- United Nations Conference on Trade and Development (UNCTAD), "Promoting Technology Adoption for Sustainable Development" [Online]. Available: https://unctad.org/system/files/official-document/tn_unctad_dtln_ipc_2019d2_en.pdf

Exercises

1. Research and discuss a government policy implemented in your country or region to promote technology adoption. Analyze its impact on technology adoption and its alignment with ethical considerations.

2. Identify a sector or industry that could benefit from technology adoption. Design a policy proposal outlining the measures that a government can implement to promote technology adoption in that sector or industry.

3. Conduct a case study on a country that has successfully implemented government policies to promote technology adoption. Analyze the key factors contributing to the success of these policies and their impact on the country's economic development.

Understanding the impact of education and training on technology adoption

Education and training play a crucial role in the adoption of technology. By equipping individuals with the necessary knowledge and skills, education and training enable them to effectively understand, use, and maximize the benefits of technological advancements. In this section, we will explore the impact of education and training on technology adoption, examining the key factors that influence the successful integration of technology into various sectors.

The importance of education in technology adoption

Education is the foundation for technology adoption. It provides individuals with the knowledge and understanding of the underlying principles and functionalities of technology. A strong educational system ensures that individuals are equipped with the necessary skills to effectively use and leverage technology for their personal and professional development.

At the individual level, education enhances digital literacy, which is essential for technology adoption. Digital literacy refers to the capability to use and navigate digital tools, devices, and platforms. It involves skills such as accessing information online, using productivity software, and communicating effectively through digital platforms. By improving digital literacy, education facilitates the adoption of technology by reducing barriers and enhancing individual confidence in using technology.

Moreover, education empowers individuals to become critical thinkers and problem solvers. It encourages creativity, innovation, and adaptability, which are essential for embracing and integrating new technologies. A strong educational foundation enables individuals to understand the potential benefits of technology and identify opportunities for its application in different sectors.

The role of training in technology adoption

While education provides the foundational knowledge, training plays a vital role in the application and integration of technology into specific contexts. Training programs focus on developing practical skills and competencies related to the use of specific technologies or tools. They enable individuals to translate their theoretical knowledge into practical applications within their respective fields.

Training programs can take various forms, including workshops, seminars, on-the-job training, and online courses. They can be designed to cater to different segments of the population, such as professionals, students, or individuals from marginalized communities. The effectiveness of training programs lies in their ability to provide hands-on experience, allowing individuals to gain confidence and competence in using technology.

Effective training programs also address the need for continuous learning and upskilling. As technology evolves rapidly, individuals must stay updated with the latest trends and developments. Training programs should, therefore, focus on providing lifelong learning opportunities to ensure that individuals can adapt to emerging technologies and remain competitive in the digital age.

Factors influencing technology adoption through education and training

Several factors influence the successful adoption of technology through education and training. These include:

1. **Accessibility of education and training:** Education and training opportunities need to be accessible and affordable for individuals from all socio-economic backgrounds. This requires the establishment of inclusive and equitable education systems that provide equal opportunities for all.

2. **Relevant curriculum:** Education programs should be designed to align with the needs of the digital age. The curriculum should include technology-related subjects and practical training that equip individuals with the skills demanded by the industry.

3. **Quality of education and training:** The quality of education and training is crucial for successful technology adoption. This involves hiring qualified

educators, providing up-to-date resources and infrastructure, and employing innovative teaching methods that promote active learning and critical thinking.

4. **Partnerships between academia and industry:** Collaboration between educational institutions and industry is essential to bridge the gap between theoretical knowledge and practical application. Partnerships can facilitate internships, apprenticeships, and research collaborations, providing students and professionals with real-world exposure to technology.

5. **Government support:** Governments play a crucial role in promoting education and training for technology adoption. They can establish policies and initiatives that incentivize educational institutions and individuals to engage in technology-related education and training.

6. **Awareness and advocacy:** Raising awareness about the benefits of technology adoption and the importance of education and training is vital. Advocacy efforts can help overcome cultural and societal barriers, ensuring that individuals see the value in embracing technology and investing in their educational development.

Example: The impact of education and training on technology adoption in healthcare

Let's consider the healthcare sector as an example to understand the impact of education and training on technology adoption. Advanced technologies, such as electronic medical record systems, telemedicine, and wearable devices, have the potential to revolutionize healthcare delivery and improve patient outcomes. However, the successful adoption of these technologies relies on the education and training of healthcare professionals.

Education equips healthcare professionals with the necessary knowledge to understand how these technologies work, their significance in patient care, and the potential benefits and risks associated with their use. Training programs enable healthcare professionals to develop the skills needed to operate and utilize these technologies effectively.

For instance, training programs can focus on teaching healthcare professionals how to navigate electronic medical record systems, interpret data generated by wearable devices, and provide telemedicine consultations. Through such training, healthcare professionals can enhance their ability to integrate technology into their clinical practice, effectively utilize available resources, and improve patient care.

Furthermore, education and training play a crucial role in fostering a culture of continuous learning and professional development within the healthcare sector. Healthcare professionals need to stay updated with the latest advancements and innovations in technology to provide quality care. By offering educational opportunities and training programs, healthcare organizations can ensure that their workforce remains competent and capable of adopting new technologies as they emerge.

Conclusion

Education and training are fundamental in facilitating the adoption of technology across various sectors. Education provides individuals with the knowledge, digital literacy, and critical thinking skills necessary to understand and leverage technology effectively. Training programs complement education by offering hands-on experience and practical skills needed for the integration of technology into specific contexts. By addressing factors such as accessibility, curriculum relevance, and quality, education and training can play a pivotal role in advancing technology adoption and driving economic growth and development.

Further Reading

1. Bonk, C. J. (Ed.). (2019). *The World is Open: How Web Technology is Revolutionizing Education.* Jossey-Bass.

2. Johnson, L., Adams Becker, S., Estrada, V., and Freeman, A. (2015). *NMC/CoSN Horizon Report: 2015 K-12 Edition.* The New Media Consortium.

3. UNESCO. (2011). *ICT Competency Framework for Teachers.* United Nations Educational, Scientific and Cultural Organization.

Exercises

1. Conduct a survey among students in your educational institution to understand their perception of the impact of education and training on technology adoption. Analyze the responses and identify common themes or challenges.

2. Research a successful case study of technology adoption in a specific industry or sector. Write a report detailing how education and training initiatives played a role in the successful integration of technology.

3. Design a training program for professionals in a specific field to improve their skills and knowledge related to a particular technology. Outline the curriculum, learning objectives, and assessment methods for the program.

Examining the role of innovation ecosystems in facilitating technology adoption

Innovation ecosystems play a crucial role in facilitating the adoption of new technologies. These ecosystems consist of various stakeholders, including entrepreneurs, startups, universities, research institutions, government agencies, and investors, who collaborate and interact to drive technological innovation and its subsequent adoption. The goal of an innovation ecosystem is to create an environment that nurtures and supports the development, diffusion, and adoption of new technologies.

Understanding the importance of innovation ecosystems

Innovation ecosystems provide a structured framework for different actors to collaborate and share resources, knowledge, and expertise. They create an environment that fosters entrepreneurship, encourages experimentation, and promotes the commercialization of innovations. By bringing together diverse stakeholders, innovation ecosystems facilitate technology adoption by addressing various challenges and leveraging synergies.

The importance of innovation ecosystems lies in their ability to:

- **Facilitate knowledge exchange and collaboration:** Innovation ecosystems bring together researchers, entrepreneurs, and experts from various disciplines, fostering interdisciplinary collaboration and knowledge sharing. This collaboration helps bridge the gap between research and practical application and encourages the exchange of ideas and expertise.

- **Provide access to funding and resources:** Innovation ecosystems help startups and entrepreneurs connect with potential investors, venture capitalists, and government funding agencies. They provide a platform for accessing financial resources, infrastructure, and specialized facilities necessary for technology adoption. This access to funding and resources is often crucial for small and medium-sized enterprises (SMEs) to develop and scale their innovations.

- **Support talent development and entrepreneurship:** Innovation ecosystems promote entrepreneurship and talent development by offering mentoring programs, training, and networking opportunities. They help entrepreneurs acquire the required skills, knowledge, and business acumen to navigate the complexities of the technology adoption process. Moreover, innovation ecosystems foster a culture of innovation, risk-taking, and learning, encouraging individuals to become entrepreneurs and adopt novel technologies.

- **Facilitate collaboration with research institutions:** Innovation ecosystems create opportunities for collaboration between startups and research institutions, such as universities and research centers. This collaboration enables startups to leverage the expertise and resources available in research institutions to develop and refine their technologies. On the other hand, research institutions benefit from real-world applications and commercialization opportunities for their innovations.

- **Promote favorable policy and regulatory environment:** Innovation ecosystems advocate for policies and regulations that support technology adoption. By engaging with government agencies and policy-makers, they can influence the regulatory framework to remove barriers and create an enabling environment for the adoption of new technologies. This may include streamlining bureaucratic processes, providing tax incentives, or establishing supportive intellectual property rights frameworks.

Analyzing the components of innovation ecosystems

Innovation ecosystems consist of several key components that collectively contribute to facilitating technology adoption. These components include:

- **Entrepreneurial support organizations:** These organizations provide essential support services to entrepreneurs and startups, such as incubators, accelerators, and co-working spaces. They offer mentorship, guidance, and access to networks, helping startups navigate the challenges of technology adoption.

- **Research and educational institutions:** Universities and research institutions contribute to innovation ecosystems by conducting cutting-edge research, developing new technologies, and nurturing talent. They often collaborate with startups and entrepreneurs, sharing their expertise and providing access to specialized facilities and resources.

- **Government agencies and policies:** Government agencies play a critical role in fostering innovation ecosystems. They provide funding, develop supportive policies, and create regulatory frameworks that enable technology adoption. Government support can significantly influence the success of innovation ecosystems by offering a stable and favorable environment for innovation.

- **Investors and funding sources:** Access to funding is crucial for startups and entrepreneurs to develop and scale their technologies. Investors, venture capitalists, and angel investors contribute to innovation ecosystems by providing financial resources and expertise. They enable entrepreneurs to access the necessary capital to commercialize their innovations.

- **Industry partnerships and collaborations:** Collaboration with industries and established companies can help startups validate their technologies, access markets, and overcome challenges related to scaling and distribution. Industry partnerships contribute to the growth and sustainability of innovation ecosystems by providing market insights, resources, and potential customers.

- **Supportive infrastructure:** Physical and digital infrastructure, such as research labs, testing facilities, and high-speed internet connectivity, are essential components of innovation ecosystems. They provide the necessary tools and resources for startups and entrepreneurs to develop and deploy their technologies.

Examining successful innovation ecosystems

Several regions and cities around the world are known for their successful innovation ecosystems. These ecosystems have played a significant role in facilitating technology adoption and driving economic growth. Let's examine two examples:

- **Silicon Valley, California, USA:** Silicon Valley is renowned for its vibrant ecosystem that has contributed to the adoption and commercialization of numerous groundbreaking technologies. The ecosystem benefits from the presence of leading universities like Stanford and UC Berkeley, research institutions like NASA Ames Research Center, and tech giants like Apple, Google, and Facebook. It has a dense network of venture capitalists, angel investors, and incubators that provide financial support and mentorship to startups. The region's culture of risk-taking, innovation, and knowledge sharing has nurtured the successful adoption of technologies across various sectors.

- **Shenzhen, China:** Shenzhen has transformed from a small fishing village to a global technology hub within a few decades. The city's innovation ecosystem is driven by close collaboration between industry, research institutions, and government agencies. It houses leading technology companies like Huawei, Tencent, and DJI, along with numerous startups and incubators. Shenzhen benefits from a favorable policy environment and government support, including tax incentives and streamlined regulations. The city's strong manufacturing capabilities, combined with access to funding and talent, have facilitated the rapid adoption and commercialization of technologies in sectors such as electronics, robotics, and advanced manufacturing.

Understanding the challenges and opportunities of innovation ecosystems

While innovation ecosystems provide numerous opportunities for technology adoption, they also face several challenges. Some of the challenges include:

- **Access to funding:** Startups and entrepreneurs often struggle to secure adequate funding for their technology adoption efforts. Limited access to financing can hinder the development and scaling of innovations. Innovation ecosystems need to address this challenge by fostering stronger connections between startups and investors, providing mentorship in fundraising, and encouraging alternative forms of financing.

- **Talent development and retention:** Developing a skilled workforce and retaining talent is crucial for the success of innovation ecosystems. Ensuring a continuous supply of skilled professionals and creating an environment that attracts and retains top talent requires investment in education, training, and professional development programs. Collaboration between industry and educational institutions can help bridge the skills gap and align educational programs with the needs of the ecosystem.

- **Regulatory barriers:** Complex and burdensome regulatory frameworks can impede the adoption of new technologies. Innovation ecosystems need to work closely with policymakers to advocate for regulations that balance consumer protection and safety with the need for technological advancements. Creating sandboxes or regulatory testbeds can provide a controlled environment for testing and validating new technologies without stifling innovation.

- **Collaboration and competition:** Balancing collaboration and competition within an innovation ecosystem can be challenging. While collaboration fosters knowledge sharing and resource-sharing, competition drives innovation and pushes entrepreneurs to excel. Innovation ecosystems must strike a balance between fostering collaboration and incentivizing competition to ensure the ecosystem's growth and sustainability.

Despite these challenges, innovation ecosystems present significant opportunities for technology adoption and economic development. By leveraging the resources, expertise, and collaboration opportunities within the ecosystem, startups and entrepreneurs can overcome obstacles and drive the widespread adoption of technologies. Governments, in collaboration with industry, academia, and other stakeholders, play a vital role in fostering an enabling environment that supports innovation ecosystems and technology adoption.

Exercises

1. Identify and analyze an innovation ecosystem in your local region. Assess its strengths, weaknesses, and the impact it has had on technology adoption and economic growth.

2. Research and discuss a case study of a successful technology adoption facilitated by an innovation ecosystem. Explain the key factors that contributed to its success and the lessons that can be learned from it.

3. Imagine you are an entrepreneur with a novel technology. Design a roadmap for navigating an innovation ecosystem to facilitate the adoption of your technology. Identify key stakeholders, resources, and strategies to overcome potential challenges.

Key Takeaways

- Innovation ecosystems play a crucial role in facilitating technology adoption by providing a collaborative and supportive environment.
- These ecosystems bring together various stakeholders, including entrepreneurs, startups, research institutions, government agencies, and investors.

- Innovation ecosystems provide access to funding, resources, expertise, and infrastructure necessary for technology adoption.
- Successful innovation ecosystems foster collaboration, talent development, and favorable policies to drive technology adoption and economic growth.
- Challenges faced by innovation ecosystems include limited access to funding, talent retention, regulatory barriers, and striking a balance between collaboration and competition.

Further Reading

1. Feld, B., & Mendelson, J. (2011). *Startup communities: Building an entrepreneurial ecosystem in your city.* John Wiley & Sons.

2. Isenberg, D. (2010). *How to start an entrepreneurial revolution.* Harvard Business Review, 88(6), 40-50.

3. OECD. (2019). *Enhancing the contributions of innovation ecosystems to productivity and inclusiveness: Policy insights from the OECD Innovation Strategy.* OECD Publishing.

4. Stam, E., Stam, C., & Audretsch, D. B. (2019). *Ambitious entrepreneurship, high-growth firms and innovation ecosystems: The European case.* Economic Policy, 34(99), 517-565.

Investigating the challenges and opportunities of technology diffusion across countries

Technology diffusion refers to the spread of technological innovations from one country to another. It plays a crucial role in economic development, as it allows countries to benefit from advancements made by others and encourages global collaboration. However, the process of technology diffusion is not without its challenges and opportunities. In this section, we will explore some of the key factors that influence technology diffusion across countries and discuss the potential benefits and drawbacks associated with it.

Factors influencing technology diffusion

Several factors influence the diffusion of technology across countries. These factors can be broadly categorized into economic, political, social, and cultural dimensions.

Let's take a closer look at each of these factors and their impact on technology diffusion:

1. **Economic factors:** Economic factors play a significant role in the diffusion of technology. In general, countries with strong economic conditions are more likely to adopt and benefit from new technologies. Factors such as income levels, investment in research and development (R&D), availability of capital, and access to financial resources greatly influence a country's ability to acquire and implement new technologies. Additionally, the presence of a skilled workforce and supportive infrastructure are also important factors in technology diffusion.

2. **Political factors:** Political stability and the presence of supportive policies and regulations are crucial for technology diffusion. Governments that prioritize technology development and provide a favorable business environment tend to attract foreign investment and encourage the adoption of new technologies. Additionally, the presence of intellectual property protection laws and well-established legal frameworks can also facilitate technology diffusion by promoting innovation and protecting the rights of inventors.

3. **Social factors:** Social factors, such as education levels, cultural attitudes towards technology, and social acceptance of change, significantly influence the diffusion of technology. Countries with well-educated populations are more likely to adopt and adapt to new technologies effectively. Additionally, cultural attitudes towards innovation and risk-taking can either facilitate or hinder technology diffusion. For example, societies that are open to change and embrace innovation tend to be more successful in adopting new technologies.

4. **Cultural factors:** Cultural factors, including language barriers, social norms, and cultural differences, can pose challenges to technology diffusion. These factors can affect the ease of communication, cooperation, and knowledge sharing between countries. Overcoming cultural barriers requires effective cross-cultural communication strategies and understanding the specific cultural contexts in which technology is being diffused.

Challenges of technology diffusion

While technology diffusion presents numerous opportunities for economic growth and development, it also brings about several challenges that must be addressed. Some of the key challenges include:

1. **Technological divide:** One of the main challenges of technology diffusion across countries is the existence of a technological divide. This divide refers to the disparities in technology access and adoption between developed and developing countries. Developing countries often face difficulties in acquiring and

implementing advanced technologies due to limited resources and infrastructure. The technological divide can exacerbate inequalities and hinder overall economic progress.

2. **Knowledge and skill gaps:** Technology diffusion demands a skilled workforce capable of understanding and utilizing new technologies effectively. However, knowledge and skill gaps often pose challenges, particularly in developing countries where educational resources may be limited. Bridging these gaps requires investments in education and training programs to develop the necessary technological expertise.

3. **Infrastructure limitations:** Adequate infrastructure is crucial for technology diffusion. However, developing countries often face challenges related to insufficient telecommunications networks, limited internet access, and inadequate transportation systems. These limitations can impede the diffusion and effective utilization of technology. Addressing infrastructure limitations requires substantial investment and long-term planning.

4. **Financial constraints:** Technology acquisition and implementation can be costly, particularly for developing countries with limited financial resources. The high costs associated with technology adoption, including investment in R&D, infrastructure development, and purchasing licenses, can pose significant challenges. Overcoming these financial constraints may require international cooperation, access to financing mechanisms, and innovative funding models.

Opportunities of technology diffusion

Despite the challenges, technology diffusion presents numerous opportunities for countries to unlock their economic potential and drive sustainable development. Some of the key opportunities include:

1. **Economic growth:** Technology diffusion can stimulate economic growth by enhancing productivity, creating new industries, and driving innovation. Access to new technologies can help countries leapfrog traditional development stages, allowing them to rapidly improve their competitiveness in the global marketplace.

2. **Knowledge sharing and collaboration:** Technology diffusion promotes knowledge sharing and collaboration between countries. Through international networks and partnerships, countries can learn from each other's experiences, share best practices, and collaborate on research and development initiatives. This exchange of knowledge can accelerate technological advancements and foster innovation.

3. **Increased efficiency:** Technology diffusion enables countries to adopt more efficient production processes, resulting in increased productivity and cost savings.

By incorporating new technologies, countries can streamline their operations, reduce waste, and improve resource utilization. These efficiency gains can contribute to overall economic development and sustainability.

4. **Social development and inclusivity:** Technology diffusion can have a positive impact on social development and inclusivity. Improved access to technology can enhance education and healthcare services, promote entrepreneurship and job creation, and empower marginalized communities. By leveraging technology, countries can work towards reducing social inequalities and promoting equitable development.

Case study: Mobile banking in Kenya

A prominent example of successful technology diffusion is the development of mobile banking in Kenya. In the early 2000s, Kenya faced challenges related to limited access to formal banking services, especially in rural areas. Traditional brick-and-mortar banks were unable to reach a significant portion of the population, leaving them unbanked or underbanked.

Recognizing the potential of technology to address this issue, Safaricom, a leading mobile network operator in Kenya, launched M-Pesa in 2007. M-Pesa is a mobile money transfer and payment service that allows users to deposit, withdraw, and transfer money using their mobile phones. The service leverages basic mobile technology to provide financial services to unbanked and underbanked individuals.

The diffusion of mobile banking in Kenya was driven by several factors. First, the widespread use of mobile phones provided a platform for delivering financial services. Second, the ease of use and convenience of mobile banking appealed to a wide range of users. Third, the partnership between Safaricom and established financial institutions ensured the security and reliability of the service.

The impact of mobile banking on Kenya's economy has been significant. It has increased financial inclusion, allowing individuals to access banking services and make digital transactions. This has provided opportunities for entrepreneurship and improved access to credit and savings. Moreover, mobile banking has facilitated the efficient transfer of remittances and reduced the reliance on cash, leading to increased security and transparency in financial transactions.

The success of mobile banking in Kenya has inspired similar initiatives across the globe, demonstrating the potential of technology diffusion in addressing societal challenges and driving economic growth.

Conclusion

Technology diffusion across countries is a complex process influenced by various economic, political, social, and cultural factors. While it presents challenges, such as the technological divide and knowledge gaps, it also offers numerous opportunities for economic growth and sustainable development. By addressing these challenges and capitalizing on the opportunities, countries can harness the power of technology to propel their economies forward. The case study of mobile banking in Kenya exemplifies the transformative impact of technology diffusion and highlights the potential for innovative solutions to drive inclusive and equitable development. It is crucial for policymakers, researchers, and stakeholders to collaborate and develop strategies that promote the responsible and effective diffusion of technology across countries. By doing so, we can unlock the full potential of technology and create a more prosperous and interconnected world.

Assessing the role of international collaborations in technology diffusion

Technology diffusion refers to the spread and adoption of new technologies across different countries and regions. It plays a crucial role in promoting economic development and driving innovation. International collaborations have become increasingly important in facilitating the process of technology diffusion, as they enable knowledge sharing, resource pooling, and collaborative research and development efforts. In this section, we will assess the key role played by international collaborations in technology diffusion and explore various strategies and examples for effective collaboration.

Understanding the importance of international collaborations

International collaborations provide numerous benefits in technology diffusion. They allow countries to leverage each other's strengths and resources, leading to faster and more widespread adoption of new technologies. Collaborative efforts can enhance research and development capabilities by combining expertise and sharing knowledge, ultimately leading to the creation of more innovative technologies. Furthermore, international collaborations can help reduce duplication of efforts and promote efficiency in the process of technology diffusion.

One of the primary benefits of international collaborations is the access to a larger and more diverse pool of research and development resources. Collaboration enables sharing of scientific knowledge, technical expertise, and infrastructure, which may otherwise be limited within a single country. By pooling resources, nations can overcome resource constraints and accelerate technological advancements.

Moreover, international collaborations facilitate the exchange of best practices and lessons learned. Different countries have unique experiences and approaches to technology adoption and diffusion. By sharing these insights, countries can learn from each other's successes and failures and develop more effective strategies for technology diffusion.

Forms of international collaborations

International collaborations can take various forms, depending on the nature of the technology and the goals of the collaboration. Some common forms of collaborations include:

- **Research partnerships:** These collaborations involve joint research projects between universities, research institutions, and industrial partners from different countries. The aim is to advance scientific knowledge and develop new technologies through shared expertise and resources. Such collaborations often result in the publication of research papers, patents, and the development of innovative solutions.

- **Technology transfer agreements:** These agreements involve the transfer of technology, know-how, or intellectual property from one country to another. Technology transfer can occur through licenses, joint ventures, or strategic alliances. These collaborations enable countries to access and adopt new technologies that may be developed or owned by another country.

- **International networks and consortia:** These collaborations involve the creation of networks or consortia consisting of multiple partners from different countries. These networks facilitate knowledge exchange, joint research, and collaborative projects. International networks provide a platform for collaboration on specific technologies, research themes, or industries. Examples include international space collaborations, such as the International Space Station, and international research networks in fields like biotechnology or nanotechnology.

- **Public-private partnerships:** These collaborations involve cooperation between government entities and private companies from different countries. The collaboration aims to combine public sector resources, policy support, and private sector expertise and funding. Public-private partnerships often focus on developing and implementing technologies with societal impact, such as sustainable energy technologies or healthcare innovations.

Successful examples of international collaborations

Numerous successful examples of international collaborations have played a pivotal role in technology diffusion. These collaborations have resulted in significant advancements in various sectors and have facilitated economic growth. Some notable examples include:

- **CERN:** The European Organization for Nuclear Research (CERN) is a prime example of an international collaboration in the field of particle physics. CERN brings together scientists from around the world to collaborate on experiments and advance our understanding of fundamental physics. The collaboration has resulted in numerous technological innovations, such as the World Wide Web, which originated at CERN and has had a transformative impact on global communication.

- **Human Genome Project:** The Human Genome Project, an international research effort to map and sequence the human genome, exemplifies the power of international collaborations in the field of biotechnology. Scientists from multiple countries joined forces to complete the project, which led to significant advancements in genomics and personalized medicine.

- **International Space Station:** The International Space Station (ISS) is a collaborative project involving multiple space agencies from different countries, including NASA, Roscosmos, ESA, JAXA, and CSA. The ISS serves as a research laboratory and a platform for scientific experiments in space. The collaboration has resulted in technological advancements in space exploration, telecommunications, and materials science, among others.

- **Clean Energy Ministerial (CEM):** The CEM is a global forum that brings together governments and private sector leaders to accelerate the transition to clean energy technologies. It facilitates international collaborations on research, policy development, and deployment of clean energy solutions. The CEM has played a key role in promoting the diffusion of renewable energy technologies worldwide.

Challenges and considerations

While international collaborations offer significant advantages in technology diffusion, they also face certain challenges and considerations. These include:

- **Intellectual property rights (IPR) protection:** Collaboration often involves the exchange of intellectual property, which raises concerns about protecting the rights of innovators and inventors. Establishing a robust framework for IPR protection is essential to encourage knowledge sharing while safeguarding the interests of collaborating parties.

- **Cultural and institutional differences:** Collaborations between countries with different cultural, political, and institutional contexts can face challenges due to varying regulatory environments, communication styles, and decision-making processes. Understanding and addressing these differences is crucial for effective collaboration.

- **Power dynamics and equitable collaborations:** Collaborations should strive for fairness and equitable participation of all partners. Power dynamics, including resource disparities and knowledge imbalances, need to be carefully managed to ensure that the benefits of collaboration are shared in an equitable manner.

- **Data sharing and security:** International collaborations often involve data sharing, which raises concerns about privacy, data protection, and cybersecurity. Establishing protocols and mechanisms for secure data sharing and ensuring compliance with relevant regulations is essential.

Conclusion

International collaborations play a crucial role in technology diffusion by enabling knowledge sharing, resource pooling, and joint research and development efforts. These collaborations provide countries with access to a larger pool of resources, facilitate the exchange of best practices, and promote efficiency in the diffusion process. Successful examples, such as CERN, the Human Genome Project, the International Space Station, and the Clean Energy Ministerial, demonstrate the transformative potential of international collaborations. However, challenges related to intellectual property rights protection, cultural differences, equitable collaborations, and data sharing need to be carefully considered and addressed. Overall, fostering international collaborations is vital to promote technology diffusion and drive economic development on a global scale.

Analyzing successful case studies of technology adoption and diffusion

In this section, we will examine several case studies of successful technology adoption and diffusion in various industries and countries. These case studies will highlight the different factors and strategies that contribute to successful implementation and widespread adoption of technology. By analyzing these real-world examples, we can gain insights into best practices and identify key drivers for successful technology adoption.

Case Study 1: Mobile Banking in Kenya

One notable case study of successful technology adoption and diffusion is the implementation of mobile banking in Kenya. In Kenya, many individuals lacked access to traditional banking services, limiting their ability to save money, access credit, and engage in financial transactions. Recognizing this gap, Safaricom, a leading telecommunications company in Kenya, launched the mobile banking service M-Pesa in 2007.

M-Pesa allowed users to deposit, withdraw, and transfer money using their mobile phones. It provided a convenient and secure alternative to traditional banking services. The success of M-Pesa can be attributed to several factors:

- **Ease of Use:** M-Pesa was designed to be simple and user-friendly, making it accessible to individuals with limited technical knowledge.

- **Mobile Network Coverage:** Safaricom had an extensive network coverage in Kenya, enabling a large population to access and use M-Pesa services.

- **Partnerships:** Safaricom collaborated with various stakeholders, including banks, merchants, and utility companies, to expand the usability of M-Pesa.

- **Trust and Security:** Safaricom implemented robust security measures to protect user data and transactions, building trust among users.

- **Government Support:** The Kenyan government embraced M-Pesa and provided an enabling regulatory environment, further encouraging its adoption.

As a result of these factors, M-Pesa achieved remarkable success. Today, it has over 40 million active users and has transformed the way Kenyans manage their finances. The success of M-Pesa has also inspired similar mobile banking initiatives in other countries, demonstrating the potential of technology adoption to drive financial inclusion.

Case Study 2: Electric Vehicles in Norway

Another compelling case study of technology adoption and diffusion is the widespread adoption of electric vehicles (EVs) in Norway. Norway has become a global leader in EV adoption, with electric cars accounting for a significant portion of new vehicle sales in the country. The success of EV adoption in Norway can be attributed to several factors:

- **Government Incentives:** The Norwegian government introduced a range of incentives to promote EV adoption, including tax exemptions, reduced tolls, and free public charging infrastructure.

- **Infrastructure Development:** Norway invested heavily in building a robust charging infrastructure network, ensuring that EV owners have convenient access to charging stations across the country.

- **Public Awareness and Education:** Public awareness campaigns highlighting the environmental and economic benefits of EVs played a crucial role in changing the public perception and promoting adoption.

- **Collaboration Between Stakeholders:** Automakers, energy companies, and government agencies collaborated to address common challenges and accelerate the adoption of EVs.

- **Long-Term Vision and Commitment:** Norway set ambitious targets for reducing greenhouse gas emissions and committed to phasing out fossil fuel vehicles, providing clear signals for the industry and consumers.

As a result of these efforts, electric vehicles now make up a significant share of the Norwegian automotive market. In 2019, EVs accounted for over 50% of new car sales in Norway. This success has positioned Norway as a model for other countries looking to transition to sustainable transportation.

Case Study 3: Cloud Computing in the Healthcare Industry

Cloud computing is another technology that has seen widespread adoption in various industries. In the healthcare sector, cloud computing offers several advantages, including cost savings, scalability, and improved collaboration. One notable case study of successful cloud computing adoption is the U.S. Department of Veterans Affairs (VA).

The VA implemented a cloud-based electronic health record (EHR) system known as VistA. The VistA system allowed healthcare providers within the VA network to access and share patient information securely. Some key factors contributing to the success of VistA adoption include:

- **Efficiency and Accessibility:** The cloud-based EHR system streamlined workflows, reduced paperwork, and ensured that patient information was accessible to authorized healthcare providers in real-time.

- **Collaboration and Coordination:** VistA improved coordination of care among different VA healthcare facilities by enabling seamless sharing of medical records and facilitating communication between providers.

- **Cost Savings:** By adopting a cloud-based system, the VA reduced infrastructure costs, eliminated the need for physical storage and backup systems, and improved overall operational efficiency.

- **Data Security and Privacy:** The VA implemented rigorous security measures to protect patient data from unauthorized access, ensuring compliance with privacy regulations.

- **Training and Support:** The successful adoption of VistA was supported by comprehensive training programs and ongoing technical support for healthcare providers.

The implementation of VistA has significantly improved the delivery of healthcare services to veterans in the U.S., demonstrating the transformative potential of cloud computing in the healthcare industry.

Conclusion

These case studies highlight the importance of various factors in driving successful technology adoption and diffusion. Factors such as ease of use, infrastructure development, government support, collaboration between stakeholders, and public awareness play a crucial role in overcoming barriers to adoption and ensuring widespread utilization of technology. By learning from these successful case studies, policymakers, businesses, and individuals can develop effective strategies for adopting and diffusing technology in their respective domains.

Understanding the ethical implications of technology adoption and diffusion

In the process of technology adoption and diffusion, it is crucial to consider the ethical implications that arise from the use and spread of new technologies. As technology continues to advance and permeate various sectors of society, it is important to critically reflect on the ethical consequences and responsibilities that come with these advancements. This section will delve into the key ethical considerations related to technology adoption and diffusion, exploring the potential benefits and risks associated with technological advancements.

The Ethical Framework

When considering the ethical implications of technology adoption and diffusion, it is helpful to establish a framework that guides decision-making and assessment of these implications. One commonly used ethical framework is the "technology ethics" approach, which involves examining the impact of technology on society from various ethical perspectives.

The technology ethics framework encompasses several ethical dimensions:

1. Privacy: Technology adoption and diffusion often involve the collection and storage of vast amounts of personal data. Ethical considerations arise regarding the protection of privacy and the responsible use of personal information.

2. Security: The rapid adoption of technology also brings concerns about cybersecurity and data breaches. The ethical responsibility lies in implementing robust security measures to protect individuals' and organizations' data from unauthorized access.

3. Equity: Technological advancements can exacerbate existing social inequalities, leading to a digital divide. Ethical considerations require efforts to bridge this divide and ensure equitable access to technology and its benefits.

4. Transparency: With the increasing complexity of technology, there is a need for transparency in the decision-making process behind technology adoption. Ethical considerations include providing clear information about the purpose, function, and potential risks associated with new technologies.

5. Accountability: As technology becomes more pervasive, the accountability of individuals and organizations involved in its adoption and diffusion becomes crucial. Ethical considerations include establishing mechanisms to hold responsible parties accountable for any harm caused by the misuse or unintended consequences of technology.

Ethical Considerations in Technology Adoption

During the process of adopting new technologies, it is essential to consider the potential ethical implications. Here are some key ethical considerations:

1. Informed Consent: When adopting new technologies, individuals and organizations must obtain informed consent from stakeholders who are affected by the technology. This includes providing clear information about the purpose, potential risks, and benefits associated with the technology, ensuring individuals can make informed decisions.

2. Privacy Protection: As technology adoption often involves the collection and use of personal data, it is crucial to prioritize privacy protection. Organizations should implement robust data protection measures, including encryption, anonymization, and strict access controls, to preserve individuals' privacy rights.

3. Data Ownership and Control: Ethical considerations center around ensuring individuals have ownership and control over their personal data. Technology adoption should respect individuals' rights to control how their data is used and shared, preventing unauthorized use or exploitation.

4. Social Implications: Adopting new technologies should consider the broader social implications. Ethical considerations include minimizing any negative social impacts, such as job displacement or inequalities, and maximizing positive impacts, such as improved accessibility and inclusivity.

Ethical Considerations in Technology Diffusion

Once a technology is adopted, its diffusion and widespread use introduce additional ethical considerations. Some key ethical considerations in technology diffusion are:

1. Equity and Access: Technology diffusion should aim to bridge the digital divide and ensure equitable access. Ethical considerations include addressing barriers to access, such as affordability, infrastructure limitations,

and technological literacy, to prevent the marginalization of certain individuals or groups.

2. Social Impact Assessment: Diffusing technology should involve conducting thorough social impact assessments to evaluate the potential consequences on different stakeholders. This assessment helps identify and mitigate any adverse effects on society, allowing for more responsible and ethical diffusion.

3. Responsible Innovation: Technological diffusion should align with principles of responsible innovation, considering the long-term social and environmental impacts. Ethical considerations include anticipating and addressing issues related to algorithmic biases, unintended consequences, and the ethical use of emerging technologies like artificial intelligence and autonomous systems.

4. Just Transition: When diffusing new technologies, particular attention must be given to managing the transition for individuals and communities affected by technological disruptions. Ethical considerations include providing support, retraining, or alternative employment opportunities to mitigate any negative social or economic impacts.

It is essential to note that ethical considerations evolve alongside advancements in technology. As technological landscapes constantly change, regular reassessment and adaptation of ethical frameworks and practices are necessary to address emerging challenges and opportunities.

Case Study: Ethical Implications of Facial Recognition Technology

To illustrate the ethical implications of technology adoption and diffusion, let's explore the case of facial recognition technology. Facial recognition technology has gained significant attention in recent years due to its numerous applications but has also raised ethical concerns.

One ethical consideration is privacy. Facial recognition technology relies on capturing and processing individuals' facial data, raising concerns about the potential misuse or unauthorized access to this sensitive information. Access limitations, strong encryption techniques, and explicit user consent can help address these concerns.

Another ethical consideration is bias and discrimination. Facial recognition systems have been shown to exhibit biases, particularly against people of color and women. These biases can lead to unjust treatment, such as misidentification or increased scrutiny of specific individuals or communities. Regular auditing, bias detection, and diverse data sets for training can help mitigate these biases.

Moreover, facial recognition technology can infringe upon personal autonomy and freedom. In public spaces, indiscriminate use of this technology may violate individuals' rights to privacy and anonymity. Clear guidelines, restrictions on deployment, and public consultation can balance the benefits and risks associated with its use.

It is crucial to engage in open and inclusive discussions about the ethical implications of facial recognition technology and establish regulations and standards to ensure responsible adoption and diffusion.

Conclusion

Understanding the ethical implications of technology adoption and diffusion is critical for promoting responsible and sustainable technological development. Privacy protection, security, equity, transparency, and accountability should be at the forefront of ethical considerations. By incorporating these considerations into decision-making processes, we can harness the benefits of technology while minimizing potential risks and adverse impacts on individuals and society at large.

Discussing strategies for fostering inclusive and responsible technology adoption

In order to ensure that the benefits of technology are accessible to all and that its adoption is done in a responsible manner, it is crucial to develop strategies that foster inclusive and responsible technology adoption. This section will explore various approaches and initiatives that can be implemented to achieve this goal.

Understanding the importance of inclusive technology adoption

Inclusive technology adoption refers to the process of ensuring that technology is accessible and beneficial to all individuals and communities, regardless of their socioeconomic status, geographic location, or any other factors that may create barriers to adoption. It recognizes the value of diversity and aims to minimize inequalities that may arise from unequal access to and use of technology.

Promoting inclusive technology adoption is essential for several reasons. Firstly, it helps to bridge the digital divide and ensure equal opportunities for all individuals to participate in the digital economy. Secondly, it reduces the risk of exclusion and marginalization of certain groups, such as the elderly, people with disabilities, and those in remote or disadvantaged areas. Finally, inclusive technology adoption promotes social cohesion and enhances overall societal well-being.

Addressing barriers to technology adoption

To foster inclusive and responsible technology adoption, it is crucial to identify and address the barriers that prevent individuals and communities from accessing and effectively utilizing technology. Some common barriers include:

1. **Infrastructure:** Limited access to reliable internet connectivity, electricity, and other essential technological infrastructure can hinder technology adoption. Efforts should be made to expand infrastructure development in underserved areas.

2. **Cost:** The high cost of technology devices and services can be a significant barrier, particularly for low-income individuals and communities. Encouraging the development of affordable technology solutions and implementing subsidy programs can help overcome this barrier.

3. **Digital literacy:** Lack of basic digital skills and knowledge can prevent individuals from fully benefiting from technology. Investing in digital literacy programs and providing training opportunities can empower individuals to adopt and use technology effectively.

4. **Usability and design:** Technology that is not user-friendly or accessible can create barriers for individuals with limited technical expertise or disabilities. Ensuring that technology is designed with inclusivity and accessibility in mind can enhance adoption rates.

5. **Language and cultural barriers:** Language and cultural differences can impact the adoption of technology in diverse communities. Efforts should be made to provide technology solutions in multiple languages and consider cultural sensitivities when designing and promoting technology.

Addressing these barriers requires a collaborative effort involving government, private sector organizations, non-profit organizations, and communities themselves. Providing financial support, incentives, and targeted programs can help overcome these barriers and promote inclusive technology adoption.

Promoting responsible technology adoption

Responsible technology adoption goes beyond accessibility and addresses the ethical, social, and environmental implications of technology implementation. It involves considering the potential risks associated with technology and ensuring that its adoption aligns with societal values and principles.

1. **Ethical considerations:** Technology adoption should be guided by ethical principles, ensuring that it respects individual privacy, data protection, and human rights. Organizations should adopt responsible data management practices and adhere to ethical standards in the development and use of technology.

2. **Sustainability and environmental impact:** Technology should be adopted in a manner that minimizes its environmental footprint and contributes to sustainable development. This can be achieved by promoting energy-efficient technologies, responsible e-waste management, and sustainable sourcing of raw materials.

3. **Stakeholder engagement:** It is essential to involve and engage all relevant stakeholders, including government, industry, civil society, and local communities, in the decision-making process regarding technology adoption. This ensures that diverse perspectives are considered, and potential risks are identified and addressed.

4. **Continuous monitoring and evaluation:** Regular monitoring and evaluation of technology adoption initiatives are vital to assess their impact and identify areas for improvement. This includes monitoring the social, economic, and environmental outcomes of technology adoption and making necessary adjustments to ensure responsible implementation.

5. **Public awareness and education:** Promoting public awareness and understanding of technology and its implications is crucial for responsible adoption. This can be done through education campaigns, public consultations, and transparent communication about the potential benefits and risks of new technologies.

By incorporating these strategies into technology adoption initiatives, it is possible to foster inclusive and responsible technology adoption. This will not only ensure that the benefits of technology are shared by all members of society but also contribute to the sustainable and equitable development of economies and communities.

Case study: Digital literacy programs in rural communities

To illustrate the impact of inclusive and responsible technology adoption strategies, let's consider the case of digital literacy programs implemented in rural communities.

In many rural areas, individuals face barriers in accessing and effectively using technology due to limited resources and digital skills. Digital literacy programs aim to bridge this gap by providing training and support for community members to develop essential digital skills.

These programs can include training sessions on basic computer skills, internet usage, online safety, and the use of specific software applications. They are often delivered through partnerships between government agencies, non-profit organizations, and local community centers.

By equipping individuals with digital literacy skills, these programs empower them to harness the benefits of technology for educational, economic, and social purposes. They can access online educational resources, develop online business ventures, and connect with others to build supportive networks.

The implementation of digital literacy programs in rural communities also highlights the importance of local engagement and collaboration. Involving community members in the design and delivery of these programs ensures that the content is relevant and addresses the specific needs and challenges faced by the community.

Overall, this case study demonstrates how fostering inclusive and responsible technology adoption through digital literacy programs can empower individuals and communities to overcome barriers and fully participate in the digital age.

Conclusion

Strategies for fostering inclusive and responsible technology adoption are essential for ensuring equitable access to and use of technology. By addressing barriers, promoting ethical considerations, and engaging stakeholders, we can create an environment where technology adoption is inclusive and beneficial for all. Through the implementation of initiatives like digital literacy programs, we can empower individuals and communities to leverage technology for their social, economic, and personal development. Ultimately, these efforts contribute to creating a more equitable and sustainable future driven by responsible technology adoption.

Technology and Entrepreneurship

Understanding the intersection of technology and entrepreneurship

In today's rapidly evolving world, the intersection of technology and entrepreneurship has become a powerhouse of economic growth and innovation. Technology has transformed the way businesses operate, creating new opportunities for entrepreneurs and driving advancements in various sectors. In this section, we will explore the crucial relationship between technology and entrepreneurship, examining how entrepreneurs leverage technology to disrupt industries, create value, and drive economic development.

The Role of Technology in Entrepreneurship

Technology plays a pivotal role in enabling entrepreneurship by providing entrepreneurs with tools, resources, and opportunities to identify and capitalize on market gaps. Here are some key aspects of the relationship between technology and entrepreneurship:

1. **Innovation Catalyst:** Technology acts as a catalyst for innovation, driving entrepreneurs to develop new products, services, and business models. Technological advancements often create a ripple effect, opening up possibilities for disruptive ideas and enabling entrepreneurs to solve complex problems more efficiently.

2. **Market Expansion:** Technology has dramatically expanded the reach of entrepreneurs beyond traditional geographical boundaries. With the advent of the internet and digital platforms, entrepreneurs can access global markets, connect with customers worldwide, and scale their businesses in ways that were previously unimaginable.

3. **Efficiency Enhancement:** Technology enhances operational efficiency, allowing entrepreneurs to streamline processes, automate tasks, and improve productivity. By leveraging technologies such as cloud computing, data analytics, and workflow automation, entrepreneurs can maximize efficiency and focus more effectively on value creation.

4. **Disruption Potential:** Technology has the power to disrupt existing industries and create new market opportunities. Entrepreneurs who leverage emerging technologies, such as artificial intelligence, blockchain, and virtual reality, can disrupt traditional business models and carve out niches in untapped markets.

5. **Collaboration Enabler:** Technology facilitates collaboration and knowledge sharing among entrepreneurs, enabling them to connect, collaborate, and learn from each other. Online communities, social media platforms, and coworking spaces have become breeding grounds for collaborative innovation, fostering a supportive ecosystem for entrepreneurial growth.

Challenges and Opportunities in Technology Entrepreneurship

While the intersection of technology and entrepreneurship holds immense potential, it also presents unique challenges and opportunities for entrepreneurs.

TECHNOLOGY AND ENTREPRENEURSHIP

Let's explore some of the key challenges and opportunities in technology entrepreneurship:

1. **Rapid Technological Advancements:** Keeping up with the pace of technological advancements can be daunting for entrepreneurs. Continuous learning and staying updated with emerging technologies are crucial to leverage their full potential. Entrepreneurs should invest in learning resources, attend technology conferences, and collaborate with experts to navigate the fast-paced technology landscape.

2. **Competition and Scalability:** As technology lowers barriers to entry, entrepreneurs face intense competition in the market. Scalability becomes a critical factor for success. Entrepreneurs need to design their businesses with scalability in mind, considering both technological scalability (ability to handle increasing demands) and business scalability (ability to enter new markets and expand operations).

3. **Cybersecurity and Data Privacy:** With increased reliance on technology, entrepreneurs face cybersecurity threats and must prioritize data privacy. Implementing robust cybersecurity measures and adhering to data privacy regulations are essential to protect sensitive information and maintain customer trust.

4. **Access to Capital:** Funding is often a significant challenge for technology entrepreneurs, as the initial investment required for research, development, and scaling can be substantial. Entrepreneurs need to explore various funding options such as venture capital, angel investors, crowdfunding, and government grants to secure the necessary capital for their ventures.

5. **Ethical Considerations:** The intersection of technology and entrepreneurship raises ethical dilemmas that entrepreneurs must navigate. They need to consider the ethical implications of their products or services, ensure responsible use of technology, and address potential social, environmental, and privacy issues.

Successful Case Studies

Examining successful case studies provides valuable insights into how technology and entrepreneurship have merged to drive innovation and economic growth. Let's explore two prominent examples:

1. **Uber:** Uber, a prime example of technology-enabled entrepreneurship, disrupted the traditional taxi industry by leveraging mobile technology and its seamless integration with transportation services. By connecting drivers and passengers through a user-friendly app, Uber transformed the way people hail and travel in urban areas worldwide, creating new market opportunities and challenging the status quo.

2. **SpaceX:** SpaceX, founded by Elon Musk, exemplifies the intersection of technology, entrepreneurship, and space exploration. By developing groundbreaking technologies and reusable rockets, SpaceX has revolutionized the space industry. Through innovative approaches and cost-effective space missions, SpaceX has disrupted the traditional space sector, opening up new possibilities for space exploration, satellite deployment, and interplanetary travel.

These case studies illustrate how technology can empower entrepreneurs to create impactful solutions, disrupt industries, and shape the future.

Future Trends and Challenges

As technology continues to advance, the intersection of technology and entrepreneurship will witness exciting future trends and face new challenges. Some trends and challenges to watch out for include:

1. **Artificial Intelligence (AI) and Automation:** AI and automation are expected to revolutionize various industries, presenting both opportunities and challenges. Entrepreneurs need to harness the power of AI while addressing concerns related to job displacement, ethics, and fairness in decision-making algorithms.

2. **Emerging Technologies:** Entrepreneurs should stay informed about emerging technologies such as blockchain, Internet of Things (IoT), augmented reality (AR), and 5G, as they have the potential to reshape industries and create new entrepreneurial opportunities.

3. **Sustainability and Social Impact:** Entrepreneurs are increasingly focusing on sustainable, socially responsible businesses. Technologies that enable clean energy, circular economy, and social impact initiatives will gain prominence, providing avenues for entrepreneurs to make a positive difference while generating economic value.

4. **Digital Divide:** Bridging the digital divide in terms of accessibility and affordability of technology is crucial for inclusive technology entrepreneurship. Entrepreneurs need to consider the needs of underserved communities, developing countries, and marginalized groups to ensure equitable access and opportunities.

In conclusion, the intersection of technology and entrepreneurship is a thriving domain with immense potential for economic growth, innovation, and societal impact. Entrepreneurs who embrace technology, adapt to emerging trends, and navigate the associated challenges can harness its transformative power to create successful ventures and drive positive change in the world. By understanding this intersection, aspiring entrepreneurs can gain valuable insights and knowledge to shape their own entrepreneurial journeys.

Analyzing the role of startups in technological innovation and economic growth

Startups play a crucial role in driving technological innovation and fueling economic growth. These young, dynamic companies are known for their disruptive ideas and ability to challenge established norms. In this section, we will explore the unique contributions of startups to technological innovation and their impact on economic development.

Importance of Startups in Technological Innovation

Startups are at the forefront of technological innovation, introducing new products, services, and business models to the market. Their ability to think outside the box and take risks enables them to develop groundbreaking solutions that address emerging challenges. Here are some key reasons why startups are vital for technological innovation:

1. **Flexibility and Agility:** Startups have the advantage of being nimble and adaptable. They can quickly respond to market demands, pivot their strategies, and experiment with new ideas. This flexibility allows them to bring innovative solutions to market faster than larger, more traditional companies.

2. **Culture of Innovation:** Startups often foster a culture of innovation where creativity and out-of-the-box thinking are encouraged. This environment promotes exploration, collaboration, and rapid iteration, leading to the development of novel technologies and solutions.

3. Focus on Disruption: Startups are disruptors by nature. They challenge established industries and business models, driving innovation and pushing the boundaries of what is possible. By questioning traditional practices, startups open up new opportunities for technological advancements.

4. Access to Emerging Technologies: Startups have the advantage of being early adopters of emerging technologies. They can leverage advancements in areas such as artificial intelligence, blockchain, and Internet of Things (IoT) to develop innovative products and services that transform industries.

Impact of Startups on Economic Growth

The growth and success of startups have far-reaching impacts on the economy. Their contribution to technological innovation translates into several economic benefits. Here are some key ways startups drive economic growth:

1. Job Creation: Startups are major engines of job creation. As they grow and expand, they create employment opportunities, attracting talent and contributing to lower unemployment rates. Startups also tend to hire younger workers and provide opportunities for skill development and entrepreneurship.

2. Increased Productivity: Startups often introduce more efficient processes, technologies, and business models, leading to increased productivity. Their innovative solutions streamline operations, decrease costs, and enable businesses to do more with less. This enhanced productivity contributes to overall economic growth.

3. Entrepreneurial Ecosystem: Startups play a vital role in nurturing entrepreneurial ecosystems. By establishing networks, mentorship programs, and funding opportunities, startups create a supportive environment for aspiring entrepreneurs. This ecosystem encourages the launch of new startups, fostering a cycle of innovation and economic development.

4. Attracting Investments: Startups with innovative technologies and growth potential attract investments from venture capitalists, angel investors, and even established companies. These investments provide startups with the necessary capital to scale their operations, develop their products, and drive further technological advancements.

5. **Competition and Market Dynamics:** Startups introduce competition into industries dominated by established players. This competition spurs innovation, lowers prices, and improves the overall quality of products and services. By challenging the status quo, startups drive market dynamics and contribute to a more vibrant and competitive economy.

Challenges Faced by Startups

While startups have immense potential, they also face numerous challenges along their journey. Understanding these challenges is essential for fostering an environment where startups can thrive. Here are some key challenges faced by startups:

1. **Access to Funding:** Startups often struggle to secure the necessary funding to fuel their growth. Traditional sources of financing may be reluctant to invest in early-stage ventures due to perceived risks. Startups need access to seed funding, venture capital, and other forms of investment to survive and scale.

2. **Talent Acquisition:** Startups require a skilled workforce to drive innovation. However, competing with larger companies for top talent can be challenging. Startups often struggle to attract and retain skilled professionals due to limited resources and competition from established firms.

3. **Regulatory and Legal Hurdles:** Startups face regulatory and legal challenges, which can slow down their progress. Compliance with industry standards, intellectual property protection, and navigating complex legal frameworks can be time-consuming and costly.

4. **Market Adoption and Scalability:** Convincing customers to adopt new technologies and solutions can be a significant hurdle for startups. Building market trust, overcoming skepticism, and demonstrating the value proposition of their offerings require both time and resources. Additionally, scaling operations to meet market demand can pose operational challenges for startups.

5. **Uncertainty and Risk:** Startups operate in a highly uncertain and risky environment, especially during the early stages. Market conditions, technological disruptions, and competitive pressures can significantly impact their prospects. Startups must navigate these uncertainties while managing limited resources and mitigating risks.

Case Study: The Role of Startups in the Tech Industry

To illustrate the impact of startups on technological innovation and economic growth, let's examine the role of startups in the tech industry. One notable example is the rise of Silicon Valley in California, USA. Silicon Valley is renowned for its concentration of startups and its contributions to the development of groundbreaking technologies.

Startups in Silicon Valley, such as Apple, Google, and Facebook, have revolutionized industries and transformed the global economy. These companies started as small ventures with innovative ideas and grew into tech giants that shape the way we live and work today. They have introduced game-changing innovations in areas like smartphones, search engines, social media, and more.

The success of startups in Silicon Valley has attracted immense investments, fostered an ecosystem of collaboration and industry support, and contributed to job creation and economic growth in the region. Silicon Valley serves as a testament to the power of startups in driving technological innovation and their profound impact on the economy.

Conclusion

Startups are essential drivers of technological innovation and economic growth. Their disruptive ideas, flexibility, and entrepreneurial culture enable them to develop novel solutions and challenge established industries. Startups play a vital role in job creation, increased productivity, attracting investments, nurturing entrepreneurial ecosystems, and fostering competition. However, startups also face significant challenges, including access to funding, talent acquisition, regulatory hurdles, market adoption, and managing uncertainty and risk. Understanding and addressing these challenges is crucial for supporting the growth and success of startups and harnessing their full potential in driving technological and economic development.

Discussing the importance of venture capital in fostering technology entrepreneurship

In the world of technology entrepreneurship, venture capital plays a crucial role in fueling innovation and driving economic growth. Venture capital firms are investment firms that provide capital to early-stage, high-growth companies in exchange for equity ownership. These firms specialize in identifying promising startups with disruptive technologies and business models, and they play a pivotal role in bridging the funding gap that often exists for entrepreneurs.

The Role of Venture Capital in Technology Entrepreneurship

Venture capital serves as a vital source of funding for technology entrepreneurs who often face difficulties securing traditional financing, such as bank loans. This is because startups are characterized by high risk, uncertain future cash flows, and a lack of tangible collateral. Venture capitalists are willing to invest in these early-stage companies because they understand the potential for exponential growth and lucrative returns on their investments.

By providing funding in the early stages, venture capital enables entrepreneurs to bring their innovative ideas to life. This funding supports research and development, product design, prototype development, market testing, and initial commercialization efforts. It also allows entrepreneurs to attract and retain top talent, develop strategic partnerships, and create brand awareness.

Moreover, venture capitalists bring much more than just financial resources to the table. They typically have deep industry knowledge and experience, valuable networks, and business acumen. This expertise can be invaluable for technology entrepreneurs who may lack the necessary business skills or experience to navigate the complexities of commercializing their ideas. Venture capitalists often take an active role in the management and strategic direction of the startups they invest in, providing guidance, mentorship, and access to key resources.

The Benefits of Venture Capital for Technology Entrepreneurs

1. Access to Capital: Venture capital provides entrepreneurs with the necessary funds to develop their technologies, hire talent, and scale their operations. This financial support is critical, especially in the early stages when startups often struggle to generate revenue or secure traditional financing.

2. Expertise and Guidance: In addition to capital, venture capitalists bring expertise, industry knowledge, and a strong network of contacts. This support can help entrepreneurs refine their business models, avoid common pitfalls, and make connections with potential customers, suppliers, and partners.

3. Validation and Credibility: Securing venture capital funding is a strong signal of validation and credibility. It demonstrates that experienced investors have confidence in the entrepreneur's vision and ability to execute. This can be crucial for attracting additional funding from other investors and building relationships with potential customers and stakeholders.

4. Long-Term Partnership: Venture capitalists typically take a long-term view of their investments, aligning their interests with those of the entrepreneurs. Unlike traditional lenders, they are willing to take on higher risks and tolerate a

longer timeline for returns. This long-term partnership provides entrepreneurs with the flexibility and runway they need to iterate, pivot, and ultimately succeed.

Challenges and Considerations in Venture Capital Investment

While venture capital offers significant benefits to technology entrepreneurs, it also poses challenges and considerations that both investors and entrepreneurs must navigate:

1. Risk and Uncertainty: Venture capital investments are inherently risky due to the high failure rate of startups. Many early-stage companies do not survive, resulting in potential loss of investment. Both investors and entrepreneurs need to carefully assess and manage these risks.

2. Dilution of Ownership: Venture capital investments often lead to dilution of ownership for entrepreneurs as new investors come on board and demand equity stakes. Entrepreneurs need to carefully consider the trade-off between financial resources and retaining control over their companies.

3. High Growth Expectations: Venture capitalists typically expect high returns on their investments. This means that entrepreneurs must be prepared for significant growth expectations, which can place pressure on management teams, require rapid scalability, and potentially shift the company's focus and priorities.

4. Alignment of Interests: It is important for entrepreneurs to align their interests and vision with those of their investors. This includes understanding the expectations, time horizon, and exit strategies of the venture capitalists. Misalignment can lead to conflicts and could hinder the growth and development of the startup.

Case Study: The Importance of Venture Capital in Technology Entrepreneurship

One notable case study that highlights the importance of venture capital in fostering technology entrepreneurship is the success story of Google. Larry Page and Sergey Brin, the founders of Google, initially received a $100,000 investment from Andy Bechtolsheim, co-founder of Sun Microsystems, in 1998. This seed investment allowed them to develop their search engine technology and lay the foundation for what would eventually become one of the most influential technology companies in the world.

As Google continued to grow, it attracted substantial venture capital investments from firms such as Kleiner Perkins and Sequoia Capital. These investments provided the necessary capital, expertise, and strategic guidance to fuel

Google's expansion and commercial success. The venture capitalists played a key role in helping Google navigate the complexities of the rapidly evolving digital landscape and make strategic decisions that propelled the company's growth.

The success of Google not only generated significant financial returns for the venture capitalists but also created a ripple effect in the technology entrepreneurship ecosystem. It inspired a new wave of entrepreneurial activity, attracted more venture capital investments into the technology sector, and contributed to the overall growth of the technology industry.

Conclusion

In conclusion, venture capital is a critical component of the technology entrepreneurship ecosystem. It provides early-stage funding, expertise, guidance, and credibility to entrepreneurs, enabling them to transform innovative ideas into successful businesses. While venture capital comes with its challenges and considerations, the potential benefits far outweigh the risks. As technology continues to drive economic growth and innovation, venture capital will remain an essential tool for fostering the development of groundbreaking technologies and fueling entrepreneurship.

Investigating the challenges and opportunities of technology entrepreneurship in developing economies

In developing economies, technology entrepreneurship presents both challenges and opportunities. On one hand, these economies often face limited resources, inadequate infrastructure, and institutional constraints. On the other hand, they can leverage technology to overcome these challenges and drive economic growth. In this section, we will explore the specific challenges and opportunities that technology entrepreneurship brings to developing economies.

Challenges of Technology Entrepreneurship

1. Limited access to capital: One of the primary challenges for technology entrepreneurs in developing economies is the limited availability of capital. Traditional sources of funding may be scarce or inaccessible due to risk aversion and uncertainty associated with technology startups. This can hinder the development and scaling of innovative ideas.

2. Weak infrastructure: Developing economies often lack basic infrastructure, such as reliable electricity, internet connectivity, and transportation systems. This

can pose significant challenges for technology entrepreneurs who rely on these services to develop and market their products or services.

3. Limited market size: Developing economies may have smaller markets compared to developed economies. This limited market size can restrict the scalability and potential revenue generation for technology startups, making it challenging to attract investment and sustain growth.

4. Lack of skilled workforce: Developing economies may face a shortage of skilled workers with the necessary technical and business expertise. This scarcity can hamper technology entrepreneurship by limiting the pool of available talent for startups.

5. Regulatory barriers: Regulatory frameworks in developing economies may not be well-suited to accommodate the rapid pace of technological innovation. Complex and cumbersome regulations can create barriers to entry and hinder the growth of technology startups.

Opportunities of Technology Entrepreneurship

1. Addressing local challenges: Developing economies often face unique challenges that can be addressed through technology entrepreneurship. By developing innovative solutions tailored to these contexts, entrepreneurs can improve the quality of life, enhance access to essential services, and address societal challenges.

2. Leapfrogging traditional barriers: Developing economies have the opportunity to leapfrog traditional development paths by adopting and leveraging new technologies. By sidestepping outdated or inefficient technologies, entrepreneurs can quickly catch up with developed economies and create new market opportunities.

3. Creating employment opportunities: Technology entrepreneurship has the potential to create substantial job opportunities in developing economies. As startups scale and expand, they can generate employment opportunities across various sectors, contributing to economic development and poverty reduction.

4. Fostering local innovation ecosystems: Technology entrepreneurship can foster the development of local innovation ecosystems. Through collaboration, knowledge-sharing, and resource pooling, entrepreneurs can create a supportive environment for startups, attracting investment and nurturing entrepreneurship culture.

5. Access to global markets: Advancements in technology, particularly the internet, have opened up global markets for technology startups in developing economies. With the right strategies, entrepreneurs can overcome geographical

barriers and reach customers worldwide, enabling them to tap into larger customer bases and scale their businesses.

Case Study: Technology Entrepreneurship in Kenya

Kenya provides an excellent example of how technology entrepreneurship has thrived in a developing economy. The country's mobile money service, M-Pesa, revolutionized financial access, particularly for the unbanked population. By leveraging mobile technology, M-Pesa enables users to send and receive money, pay bills, and access financial services using their mobile phones. This innovation had a significant impact on financial inclusion and economic development in Kenya.

M-Pesa's success story demonstrates how technology entrepreneurship in developing economies can address local challenges, create employment opportunities, and foster collaboration between different sectors. It also highlights the need for supportive government policies, investment in infrastructure, and the development of a vibrant startup ecosystem.

Key Strategies for Success

To harness the potential of technology entrepreneurship in developing economies, several strategies can be followed:

1. Creating conducive policy environments: Governments can play a crucial role in fostering technology entrepreneurship by implementing supportive policies and regulatory frameworks. This includes providing tax incentives, simplifying business registration processes, and protecting intellectual property rights.

2. Building technological infrastructure: Investing in technological infrastructure, such as broadband connectivity, power supply, and research facilities, is vital for supporting technology startups. Governments and stakeholders should collaborate to ensure the availability of reliable infrastructure necessary for entrepreneurship to thrive.

3. Encouraging entrepreneurship education and skills development: Developing economies should prioritize entrepreneurship education and skills development programs. This can equip aspiring entrepreneurs with the necessary knowledge, skills, and mindset to navigate the challenges of technology entrepreneurship.

4. Promoting collaboration and networking: Building strong networks and fostering collaboration among entrepreneurs, investors, academia, and government agencies is essential for the growth of technology entrepreneurship. This

collaboration can drive innovation, provide mentorship, and create access to funding opportunities.

5. Leveraging public-private partnerships: Governments and private sector entities can join forces through public-private partnerships to create a conducive environment for technology entrepreneurship. This can involve the provision of funding, mentorship, infrastructure, and other resources necessary for startups to thrive.

Conclusion

In conclusion, technology entrepreneurship in developing economies presents both challenges and opportunities. While limited access to capital, weak infrastructure, and regulatory barriers pose challenges, addressing local challenges, leapfrogging traditional barriers, creating employment opportunities, fostering local innovation ecosystems, and accessing global markets provide unique opportunities. By implementing key strategies, such as creating supportive policy environments, investing in technological infrastructure, promoting entrepreneurship education, and fostering collaboration, developing economies can unlock the full potential of technology entrepreneurship and spur economic growth and development. Entrepreneurs, governments, and stakeholders must work together to create an enabling environment that encourages innovation, supports startups, and harnesses the transformative power of technology.

Assessing the role of incubators and accelerators in supporting technology startups

Incubators and accelerators play a crucial role in supporting technology startups by providing them with the necessary resources, support, and mentorship to thrive and grow. In this section, we will explore the key functions and benefits of incubators and accelerators, discuss their impact on technology entrepreneurship, and examine the challenges and opportunities associated with their role in the startup ecosystem.

Introduction to Incubators and Accelerators

Incubators and accelerators are organizations that nurture and support early-stage startups, offering a range of resources and services tailored to the needs of these companies. While there are similarities between them, there are also key differences in their goals, structure, and focus.

Incubators typically provide long-term support and resources to startups in their early stages, helping them develop their ideas and business models, secure funding, and refine their products or services. Incubators often offer a physical space where startups can work, access to mentorship and expertise, networking opportunities, and various support services such as legal and accounting assistance.

Accelerators, on the other hand, are typically short-term programs that focus on accelerating the growth of startups. They provide a more intensive and structured environment, usually lasting for a few months, during which startups receive mentorship, training, access to investors, and resources to help them scale their business quickly. Accelerators often culminate in a demo day or investor pitch event, where startups have the opportunity to showcase their progress and attract funding.

Functions and Benefits of Incubators and Accelerators

1. **Access to Resources:** One of the primary functions of incubators and accelerators is to provide startups with access to resources that they may not have otherwise. This includes physical office space, high-speed internet, meeting rooms, and access to equipment or specialized facilities. These resources help startups reduce their initial costs and create a conducive working environment.

2. **Mentorship and Expertise:** Incubators and accelerators typically have a network of experienced mentors and industry experts who provide guidance and advice to startup founders. This mentorship helps startups navigate challenges, refine their business strategies, and make informed decisions. Mentors may also provide valuable connections to potential customers, partners, and investors.

3. **Investor Connections:** Incubators and accelerators often have strong relationships with venture capitalists, angel investors, and other funding sources. They can facilitate introductions and pitch opportunities for startups seeking investment. This access to investors greatly increases the chances of startups securing funding to fuel their growth and development.

4. **Networking Opportunities:** Joining an incubator or accelerator exposes startups to a network of like-minded entrepreneurs, industry professionals, and potential business partners. This networking can lead to collaborations, partnerships, and valuable connections that can help startups scale their operations and expand their reach.

5. **Business Development Support:** Incubators and accelerators provide startups with business development support, including help with market research, customer acquisition strategies, product development, and market validation. This support helps startups refine their value proposition, identify their target market, and develop a strong business model.

Impact of Incubators and Accelerators on Technology Entrepreneurship

Incubators and accelerators have had a significant impact on technology entrepreneurship and the startup ecosystem. They have played a crucial role in enabling the success of many innovative companies and fostering economic growth. Here are some key impacts:

1. **Increased Success Rate:** Startups that join incubators or accelerators have been found to have a higher success rate compared to those that do not. The ecosystem of support and mentorship provided by these programs helps startups overcome challenges and increase their chances of survival and growth.

2. **Access to Capital:** Incubators and accelerators provide startups with exposure to investors, increasing their access to funding. Startups that go through these programs often have higher investment readiness and are better equipped to attract investors, leading to increased funding opportunities.

3. **Job Creation:** Successful startups supported by incubators and accelerators often go on to create jobs, stimulating economic growth and employment opportunities. This job creation contributes to the overall development of local economies and helps cultivate a culture of innovation and entrepreneurship.

4. **Knowledge Spillover:** The collaborative environment within incubators and accelerators facilitates knowledge spillover among the startups and mentors. Startups can learn from each other's experiences, share best practices, and gain insights into the challenges and opportunities in the market. This collective learning fosters innovation and can lead to the development of new technologies and solutions.

Challenges and Opportunities

While incubators and accelerators offer significant benefits, they also face challenges in supporting technology startups effectively. It is important to be aware of these challenges and explore ways to address them. Some of the common challenges include:

1. **Selection and Screening:** Incubators and accelerators often need to select startups based on their potential for success. This process can be challenging as it requires effective screening mechanisms and the ability to identify startups with high growth potential. Developing thorough evaluation criteria and leveraging experienced mentors can help address this challenge.

2. **Sustainability and Funding:** Incubators and accelerators themselves need sustainable funding sources to continue providing support to startups. Overreliance on government funding or corporate partnerships can be risky.

Diversifying funding sources, developing revenue-generating programs, and building strong alumni networks can improve sustainability.

3. **Tailored Support:** Each startup is unique, with different needs and challenges. It can be challenging for incubators and accelerators to provide tailored support to each startup within a structured program. Regular feedback mechanisms, personalized mentoring, and flexible program structures can address this challenge.

4. **Post-program Support:** After completing an incubator or accelerator program, startups need continued support to sustain their growth trajectory. Ensuring that startups have access to ongoing mentorship, investor connections, and networking opportunities is crucial for their long-term success.

Case Study: Y Combinator

Y Combinator is one of the most renowned and successful accelerators globally, known for its rigorous program and successful portfolio of startups. Founded in 2005, Y Combinator has supported several technology startups, including Airbnb, Dropbox, Reddit, and DoorDash.

Y Combinator offers a three-month long accelerator program that includes seed funding, mentorship, and access to a network of successful entrepreneurs and investors. During the program, startups collaborate with mentors to refine their product, test their market fit, and present their progress at a demo day to a select group of investors.

The success of Y Combinator can be attributed to its focus on high-potential startups, its network of experienced mentors and alumni, and its emphasis on providing both financial and non-financial support. The accelerator's reputation and track record also contribute to attracting top talent and securing investments for its portfolio companies.

Conclusion

Incubators and accelerators play a vital role in supporting technology startups by providing resources, mentorship, and access to investors. Their impact on technology entrepreneurship is reflected in the increased success rates of startups that go through these programs. However, challenges such as selection and screening, sustainability, tailored support, and post-program support should be addressed to maximize their effectiveness. Case studies like Y Combinator demonstrate the positive outcomes that can be achieved through well-designed and executed incubator and accelerator programs. By fostering a supportive ecosystem for startups, incubators, and accelerators contribute to the growth of technology-driven economies and promote innovation and economic development.

Examining the Impact of Intellectual Property Rights on Technology Entrepreneurship

Intellectual property (IP) refers to the legal rights granted to individuals or organizations for their inventions, creations, or designs. It plays a crucial role in technology entrepreneurship by providing a framework for protecting and incentivizing innovation. In this section, we will examine the impact of intellectual property rights on technology entrepreneurship, considering both the benefits and challenges associated with IP protection.

Understanding Intellectual Property Rights

Intellectual property rights encompass various forms of protection, including patents, trademarks, copyrights, and trade secrets. Each form of IP protection serves a different purpose in the context of technology entrepreneurship.

- **Patents:** Patents grant inventors exclusive rights to their inventions, preventing others from making, using, or selling the patented technology without permission. They encourage innovation by providing a temporary monopoly, typically lasting 20 years, during which the inventor can commercialize the invention and gain a competitive advantage.

- **Trademarks:** Trademarks protect the distinct identity of brands, logos, or symbols associated with goods or services. They ensure that consumers can differentiate between products in the market and help businesses build a reputation and establish customer loyalty.

- **Copyrights:** Copyrights safeguard original creative works, such as literature, music, or software. They give authors or creators the exclusive right to reproduce, distribute, or display their works. Copyright protection encourages the creation and dissemination of artistic and intellectual expressions.

- **Trade Secrets:** Trade secrets include proprietary information, such as formulas, processes, or customer lists, which provide a competitive advantage. Unlike patents or copyrights, trade secrets can be protected indefinitely as long as they remain confidential.

The Role of Intellectual Property in Technology Entrepreneurship

The presence of a strong and well-enforced intellectual property regime is crucial for technology entrepreneurship. Intellectual property rights provide several key benefits for entrepreneurs:

1. **Incentivizing Innovation:** Intellectual property rights create a legal framework that encourages individuals and organizations to invest in research and development. By providing exclusivity and financial incentives, IP protection rewards innovators for their efforts and allows them to capture the economic benefits of their inventions.

2. **Attracting Investment:** Intellectual property rights increase the attractiveness of technology startups to investors. Patents, trademarks, or copyrights act as tangible assets that can secure funding and help startups establish a competitive edge in the market. Investors are more likely to support ventures that have strong IP protection.

3. **Facilitating Technology Transfer:** Intellectual property rights enable technology transfer from research institutions and universities to the commercial sector. Researchers and academics can protect their inventions through patents or copyrights, allowing them to collaborate with entrepreneurs and industry partners for further development and commercialization.

4. **Promoting Collaboration and Licensing:** Intellectual property rights facilitate collaborations and licensing agreements between technology entrepreneurs and other stakeholders. Licensing IP assets to other companies can generate revenue streams for startups and foster innovation through knowledge sharing and joint ventures.

5. **Protecting Market Position:** Intellectual property rights provide entrepreneurs with a means to protect their market position and prevent competitors from exploiting their innovations. This exclusivity allows entrepreneurs to establish a solid foothold in the market, attract customers, and build brand recognition.

Challenges in Intellectual Property Protection

While intellectual property rights offer significant benefits for technology entrepreneurship, they also pose challenges that entrepreneurs must navigate:

1. **Costs and Complexity:** Obtaining and maintaining intellectual property rights can be costly and complex. The process of filing patents or registering trademarks requires legal expertise and financial resources. For technology startups with limited budgets, this can be a barrier to IP protection.

2. **Enforcement Issues:** Intellectual property rights are only effective if they can be enforced. Enforcement can be challenging, especially in international markets where IP infringement may be difficult to detect and address. Entrepreneurs must carefully consider the jurisdictions in which they seek protection and develop strategies to enforce their rights.

3. **Intellectual Property Theft:** Technology entrepreneurs face the risk of intellectual property theft, particularly in highly competitive industries. Trade secrets or innovative technologies can be vulnerable to misappropriation or unauthorized use. Implementing robust security measures and confidentiality agreements becomes essential to protect valuable IP assets.

4. **Intellectual Property Trolls:** Patent trolls, also known as non-practicing entities, acquire and hold patents solely for the purpose of extracting licensing fees or filing infringement lawsuits. Technology entrepreneurs may face challenges when navigating legal disputes with patent trolls, diverting resources and hindering innovation.

5. **Balance with Open Innovation:** Intellectual property rights may create tensions in open innovation ecosystems, where collaboration and knowledge sharing are essential. Startups must strike a balance between protecting IP assets and engaging in open innovation practices that can catalyze technology development and market growth.

Case Study: Intellectual Property in the Technology Startup Ecosystem

To illustrate the impact of intellectual property rights on technology entrepreneurship, let's examine the case of a technology startup called InnovateTech. InnovateTech has developed a revolutionary software solution for optimizing supply chain management in the e-commerce industry. The company wants to understand the implications of intellectual property protection on its business strategy.

InnovateTech decides to file a patent for its software, protecting its novel algorithms and unique features. This move not only grants the company exclusive rights to its invention but also allows InnovateTech to attract potential investors who recognize the value of its IP assets. With the secured funding, InnovateTech can further develop its software, expand its market reach, and establish a strong market position.

However, InnovateTech also faces challenges in enforcing its patent rights, as competitors attempt to imitate its technology. The company invests in monitoring and detection mechanisms to identify instances of infringement and takes legal action against infringers. By actively protecting its intellectual property, InnovateTech demonstrates its commitment to preserving its market advantage and sustaining its growth in the competitive marketplace.

Conclusion

In conclusion, intellectual property rights play a critical role in technology entrepreneurship. They incentivize innovation, attract investment, facilitate technology transfer, promote collaboration, and protect the market position of technology startups. However, entrepreneurs must also navigate the challenges associated with IP protection, such as cost, enforcement issues, theft, patent trolls, and finding a balance between IP protection and open innovation. By understanding and effectively managing intellectual property rights, technology entrepreneurs can maximize the value of their innovations and contribute to economic growth and societal development.

According to recent studies, technology entrepreneurship plays a crucial role in driving innovation and economic growth. It refers to the process of creating, developing, and commercializing new technologies and innovations into viable business ventures. However, the success of technology entrepreneurship heavily depends on the enabling environment created by government policies. In this section, we will delve into understanding the role of government policies in promoting technology entrepreneurship.

Government Policies and Technology Entrepreneurship

Government policies have a significant impact on the ecosystem in which technology entrepreneurship operates. The right set of policies can foster an environment conducive to innovation, while inappropriate policies can impede technological advancements and hinder entrepreneurship. Let's explore some of the key policy areas that governments can focus on to promote technology entrepreneurship:

Regulatory Framework

A supportive regulatory framework is essential for technology entrepreneurship. It should provide a balance between ensuring consumer protection, maintaining fair competition, and facilitating the growth of innovative startups. The following measures can be adopted:

- **Simplified Business Registration:** Governments can streamline the process of registering a new technology venture, reducing bureaucratic hurdles and paperwork. This facilitates quick and efficient establishment of startups.

- **Intellectual Property Rights (IPR) Protection:** Strong IPR protection is crucial for encouraging technology entrepreneurship. Governments should enforce patent laws, copyrights, and trademarks to provide innovators with incentives to invest in research and development.

- **Flexible Regulations for Emerging Technologies:** Governments need to develop flexible regulations that can accommodate the rapid pace of technological advancements. This allows entrepreneurs to experiment and innovate without being stifled by outdated rules.

- **Data Privacy and Security:** Robust data privacy and security regulations are essential to build trust among consumers and protect sensitive information. Governments can establish comprehensive frameworks to regulate the collection, storage, and usage of personal and business data.

Access to Funding

Access to funding is a critical factor in the success of technology entrepreneurship. Startups often face challenges in securing the necessary capital for research, development, and scaling up their businesses. Governments can play a vital role in facilitating funding opportunities through the following mechanisms:

- **Seed Funding and Grants:** Governments can provide seed funding and grants to early-stage technology startups. This helps entrepreneurs overcome the initial financial challenges and encourages them to embark on their entrepreneurial journey.

- **Venture Capital (VC) Support:** Governments can establish partnerships with venture capital firms to provide entrepreneurs with access to larger funding pools. This can be done through co-investment programs or providing tax incentives to venture capitalists investing in technology startups.

- **Crowdfunding Platforms:** Governments can promote and regulate crowdfunding platforms that allow entrepreneurs to raise funds from a large number of individuals or organizations. Crowdfunding provides an alternative source of funding, especially for projects that may not fit traditional financing models.

Education and Skills Development

Building a skilled workforce is essential for technology entrepreneurship. Governments can focus on the following areas to foster human capital development:

- **STEM Education:** Governments should prioritize Science, Technology, Engineering, and Mathematics (STEM) education at all levels. This will equip students with the necessary knowledge and skills to pursue entrepreneurial ventures in technology-based fields.

- **Entrepreneurship Education and Training:** Integrating entrepreneurship education and training programs into the curriculum can help aspiring technology entrepreneurs develop the necessary business acumen and entrepreneurial mindset.

- **Continuing Professional Development:** Governments can encourage continuous learning among professionals in technology-related fields. This can be achieved by subsidizing training programs, organizing workshops, and providing resources for upskilling and reskilling.

Infrastructure and Support Ecosystem

A robust infrastructure and support ecosystem are critical for technology entrepreneurship to thrive. Governments can play a crucial role in developing and maintaining the necessary infrastructure and fostering a supportive ecosystem through the following measures:

- **Incubation and Acceleration Programs:** Governments can establish and support technology-focused incubators and accelerators. These programs provide startups with mentorship, networking opportunities, and access to resources to accelerate their growth.

- **Collaboration Platforms:** Governments can facilitate the creation of collaboration platforms that bring together entrepreneurs, researchers, industry experts, and investors. Such platforms encourage knowledge sharing, collaboration, and cross-pollination of ideas.

- **Access to Research Facilities:** Governments can provide technology startups with access to research facilities, laboratories, and testing centers. This enables entrepreneurs to conduct experiments, prototype development, and product testing, which may otherwise be costly for early-stage startups.

- **Public Procurement Policies:** Governments can adopt public procurement policies that encourage the adoption of innovative technologies by government agencies. This creates a market for technology startups, boosts their credibility, and provides them with early customers.

International Collaboration and Support

In an increasingly interconnected world, international collaboration and support can play a significant role in promoting technology entrepreneurship. Governments can facilitate collaboration and support through the following initiatives:

- **Global Entrepreneurship Networks:** Governments can participate in global entrepreneurship networks and initiatives to exchange knowledge, best practices, and experiences in promoting technology entrepreneurship. This helps create a supportive global ecosystem for entrepreneurs.

- **International Funding Programs:** Governments can collaborate with international organizations and funding agencies to provide technology startups with access to global funding opportunities beyond their domestic markets. This expands the pool of potential investors and partners for entrepreneurs.

- **Technology Transfer and Licensing:** Governments can establish technology transfer and licensing programs to facilitate the commercialization of research and development outputs from academic institutions and research organizations. This enables entrepreneurs to leverage existing technologies and knowledge.

Case Study: StartUp India Initiative

The StartUp India initiative launched by the Government of India provides an excellent example of comprehensive government policies aimed at promoting technology entrepreneurship. The initiative focuses on the following key areas:

- **Simplified Business Registration:** StartUp India introduced an online platform that enables entrepreneurs to register their businesses quickly and efficiently, eliminating the need for physical paperwork and reducing the time required to set up a startup.

- **Funding Support:** The initiative established a fund of funds with a corpus of INR 10,000 crores (approximately USD 1.3 billion) to provide venture capital support to startups. The government also provides credit guarantee schemes to facilitate easier access to loans for startups.

- **Tax Benefits and Incentives:** Startups registered under the initiative are eligible for tax benefits for a specified period, including exemptions from income tax and capital gains tax. The government also relaxed labor and environmental compliance for startups to reduce regulatory burdens.

- **Incubation and Acceleration Support:** StartUp India supports the establishment of incubators and accelerators across the country. These programs provide startups with mentoring, workspace, networking opportunities, and access to investors and industry experts.

- **Intellectual Property Rights (IPR) Support:** The initiative offers a fast-track mechanism for recognizing and protecting intellectual property rights for startups. This promotes innovation and encourages startups to invest in research and development.

The StartUp India initiative has played a crucial role in promoting technology entrepreneurship in India. It has created a supportive ecosystem for startups, encouraged innovation, and attracted domestic and foreign investments.

Conclusion

Government policies have a significant impact on technology entrepreneurship, either by facilitating or hindering its growth. By focusing on areas such as regulatory frameworks, access to funding, education and skills development, infrastructure and support ecosystem, and international collaboration, governments can create an enabling environment for technology entrepreneurs to thrive. Case studies like the StartUp India initiative highlight the positive impact comprehensive government policies can have on technology entrepreneurship. It is crucial for governments worldwide to understand and prioritize the role of policies in promoting technology entrepreneurship to foster innovation, economic growth, and societal development.

Key Takeaways

- Government policies play a crucial role in promoting technology entrepreneurship.
- A supportive regulatory framework, access to funding, education and skills development, infrastructure and support ecosystem, and international collaboration are key areas where governments can focus their efforts.
- Case studies such as the StartUp India initiative demonstrate the positive impact of comprehensive government policies on technology entrepreneurship.
- It is essential for governments to understand the significance of policies in creating an enabling environment for technology entrepreneurs to thrive.

Analyzing successful case studies of technology entrepreneurship

In this section, we will examine several successful case studies of technology entrepreneurship. These examples will highlight the key factors that contributed to the success of these ventures and provide insights into the strategies and approaches adopted by successful technology entrepreneurs.

1. **Apple Inc.**

Apple Inc. is a prime example of a successful technology entrepreneur. Founded by Steve Jobs, Steve Wozniak, and Ronald Wayne in 1976, the company started with the creation of personal computers. However, it was the introduction of innovative products like the iPod, iPhone, and iPad that propelled Apple to new heights.

Key factors for success:

- Innovative product design: Apple's success can be attributed to its focus on creating aesthetically pleasing and user-friendly products. The company prioritized design and simplicity, which set them apart from competitors.

- Ecosystem integration: Apple leveraged the power of its ecosystem by integrating hardware, software, and services seamlessly. This approach created a cohesive user experience and locked customers into the Apple ecosystem.

- Marketing and branding: Apple's marketing campaigns were centered around creating a brand image that resonated with consumers. Their "Think Different" campaign and iconic product launches generated significant hype and created a loyal customer base.

2. **Google**

Google is another technology entrepreneurship success story. Larry Page and Sergey Brin founded the company in 1998 as a search engine. Since then, Google has expanded its portfolio to include various products and services, such as Gmail, Google Maps, Android, and Google Cloud.

Key factors for success:

- Focus on user experience: Google prioritized delivering a seamless and relevant search experience to users. Their PageRank algorithm revolutionized web search and provided more accurate and useful search results.

- Data-driven decision-making: Google's success can be attributed to its extensive use of data analytics. The company collects and analyzes vast amounts of user data to improve its products and deliver personalized experiences.

- Continuous innovation: Google has a culture of innovation, with employees encouraged to work on passion projects and explore new ideas. This approach has led to the development of groundbreaking technologies like self-driving cars (Waymo) and artificial intelligence (Google AI).

3. **Alibaba Group**

Alibaba Group, founded by Jack Ma in 1999, is a multinational conglomerate specializing in e-commerce, retail, internet, and technology. It is one of the world's largest e-commerce companies and has expanded its business to include various sectors, including cloud computing, digital payments, and logistics.

Key factors for success:

- Market positioning: Alibaba recognized the potential of the Chinese market early on and strategically positioned itself as a platform for connecting buyers and sellers. This approach helped them capture a significant share of the rapidly growing e-commerce market in China.

- Ecosystem development: Alibaba's ecosystem approach involved building a suite of interconnected services, including Tmall, Taobao, Alipay, and AliExpress. This ecosystem created a seamless experience for users and offered a wide range of services and products.

- Adaptability and diversification: Alibaba demonstrated adaptability by expanding its business into new sectors, such as cloud computing and digital payments. This diversification helped the company navigate changing market trends and stay ahead of the competition.

These case studies provide valuable insights into the key factors that contribute to the success of technology entrepreneurship ventures. However, it is important to note that each case is unique, and success is influenced by a combination of factors, including market conditions, leadership, innovation, and timing.

To apply these lessons to real-world scenarios, consider the following exercise:

Exercise 3.2.8: Select a successful technology entrepreneurship venture in your local area or industry and analyze the key factors that contributed to its success. Compare and contrast these factors with the case studies discussed in this section. What lessons can be learned from these successful ventures?

Remember, successful technology entrepreneurship is not guaranteed, and every venture comes with its own set of challenges. However, studying successful case studies can provide valuable insights and inspiration for aspiring technology entrepreneurs.

Exploring the potential of social entrepreneurship in leveraging technology for social impact

In recent years, there has been a growing recognition of the power of social entrepreneurship in driving positive social change. Social entrepreneurs are individuals who combine entrepreneurial skills with a mission to address social or environmental challenges. They strive to create innovative and sustainable solutions that can have a lasting impact on society. With the rapid advancement of technology, social entrepreneurship has gained a new dimension, allowing for even greater potential to create social impact. In this section, we will explore the potential of social entrepreneurship in leveraging technology for social impact.

The Role of Technology in Social Entrepreneurship

Technology plays a vital role in enabling and enhancing the impact of social entrepreneurship. It provides the tools and platforms that empower social entrepreneurs to reach a wider audience, scale their initiatives, and create meaningful change. Here are some key ways in which technology can support social entrepreneurship:

1. **Access to Information and Resources:** Technology has made information and resources more accessible than ever before. Social entrepreneurs can leverage the internet and digital platforms to access knowledge, research, and best practices in their respective fields. They can also tap into online communities and networks for collaboration, mentorship, and support.

2. **Efficient Operations and Processes:** Technology offers various digital tools and applications that can streamline operations and processes for social entrepreneurs. This includes project management software, financial management tools, communication platforms, and data analytics solutions. By leveraging technology, social entrepreneurs can optimize their resources and focus more on driving social impact.

3. **Enhanced Communication and Networking:** The advent of social media and digital communication platforms has revolutionized the way we connect and interact with others. Social entrepreneurs can leverage these platforms to raise awareness about their initiatives, engage with stakeholders, and build communities of supporters. Technology enables social entrepreneurs to amplify their voice and reach a global audience.

4. **Data-driven Decision Making:** Technology allows social entrepreneurs to collect, analyze, and interpret data to make informed decisions. They can utilize data analytics tools to better understand the needs and preferences of their target beneficiaries. By harnessing the power of data, social entrepreneurs can design and implement interventions that are evidence-based and have a higher likelihood of success.

5. **Innovative Solutions:** Technology opens up new possibilities for developing innovative solutions to social challenges. Social entrepreneurs can leverage emerging technologies such as blockchain, artificial intelligence, Internet of Things (IoT), and virtual reality to create transformative interventions. For example, blockchain technology can enable transparent and accountable supply chains for fair trade products, while AI and IoT can support smart agriculture systems for sustainable food production.

Case Studies: Leveraging Technology for Social Impact

Let's explore two case studies that highlight the potential of social entrepreneurship in leveraging technology for social impact:

Case Study 1: Solar Sister

Solar Sister is a social enterprise that aims to eradicate energy poverty in rural Africa by empowering women entrepreneurs. By leveraging technology, Solar Sister has created a unique model that combines renewable energy solutions with women's economic empowerment. Solar Sister provides women with access to clean energy products such as solar lanterns, solar home systems, and clean cookstoves. These products are sold by women entrepreneurs, called Solar Sisters, who are trained and supported by the organization.

Technology plays a crucial role in the success of Solar Sister. The organization utilizes mobile technology to connect with and train Solar Sisters remotely. Solar Sisters use mobile phones to receive product orders, access marketing materials, and communicate with customers. This technology-driven approach allows Solar Sister to scale its operations efficiently and reach more marginalized communities.

Case Study 2: Health[e]Foundation

Health[e]Foundation is a social enterprise that focuses on improving healthcare delivery and education in low-resource settings. They leverage technology to provide healthcare professionals with accessible and evidence-based training programs. Health[e]Foundation offers e-learning platforms and mobile applications that deliver interactive and multimedia-rich content. These platforms cover a range of healthcare topics, including infectious diseases, maternal and child health, and non-communicable diseases.

By leveraging technology, Health[e]Foundation can reach healthcare professionals in remote and underserved areas, where access to traditional training programs is limited. The organization's digital platforms provide healthcare professionals with up-to-date knowledge and skills, enabling them to deliver quality care and contribute to improved health outcomes.

Challenges and Opportunities

While technology presents immense opportunities for social entrepreneurship, it also comes with its own set of challenges. Here are some key challenges and opportunities in leveraging technology for social impact:

1. **Accessibility and Inclusion:** The digital divide remains a significant challenge in leveraging technology for social impact. Not everyone has access to the internet, computers, or smartphones. Social entrepreneurs need to ensure that their technological solutions are accessible and inclusive, especially for marginalized communities.

2. **Digital Literacy and Skills:** To benefit from technology, individuals and communities need digital literacy and skills. Social entrepreneurs can play a role in bridging the digital skills gap by providing training and capacity-building programs. It is essential to empower individuals to effectively use technology for their socio-economic development.

3. **Financial Sustainability:** While technology can enhance efficiency and scalability, it also requires financial resources for implementation and maintenance. Social entrepreneurs need to develop sustainable business models that generate income to support their technological initiatives. This may involve partnerships, grants, impact investment, or revenue-generating activities.

4. **Ethical Considerations:** Technology raises ethical considerations in terms of privacy, data security, and algorithmic biases. Social entrepreneurs need to be mindful of these ethical implications and adopt responsible practices in their technological solutions. They should prioritize the protection of beneficiaries' rights and ensure transparency and accountability in their use of technology.

Conclusion

The potential of social entrepreneurship in leveraging technology for social impact is vast. Through the strategic use of technology, social entrepreneurs can create innovative and scalable solutions that address pressing social and environmental challenges. However, it is crucial to recognize and address the challenges associated with technology adoption and ensure inclusivity and ethical practices. As technology continues to advance, social entrepreneurship has the power to drive positive change and create a more sustainable and equitable future.

Discussing the future trends and challenges in technology entrepreneurship

The field of technology entrepreneurship is rapidly evolving, driven by continuous advancements in various technologies. In this section, we will explore some of the future trends and challenges that are likely to shape the landscape of technology entrepreneurship.

Trend 1: Artificial Intelligence (AI) and Machine Learning

One of the most significant trends in technology entrepreneurship is the increasing integration of artificial intelligence (AI) and machine learning (ML) into various business processes. AI and ML have the potential to revolutionize industries by automating routine tasks, improving decision-making, and enhancing customer experiences.

Entrepreneurs need to stay abreast of the latest AI and ML techniques and tools. Understanding how to harness the power of these technologies can provide a competitive advantage and open up new opportunities for innovation. Moreover, ethical considerations surrounding AI, such as bias and privacy, need to be addressed to ensure responsible and inclusive use of these technologies.

Trend 2: Internet of Things (IoT)

The Internet of Things (IoT) is another trend that is driving technology entrepreneurship. IoT refers to the network of interconnected physical devices, vehicles, and other objects embedded with sensors, software, and connectivity, enabling them to collect and exchange data.

Entrepreneurs can leverage IoT to create innovative products and services that facilitate automation, data analytics, and real-time decision-making. For example, IoT-enabled devices can be used to optimize energy consumption, improve supply chain management, and enhance healthcare delivery.

However, with the proliferation of IoT devices comes the challenge of ensuring robust cybersecurity measures. Entrepreneurs must prioritize cybersecurity to protect sensitive data and prevent unauthorized access to IoT networks.

Trend 3: Blockchain Technology

Blockchain technology, which underpins cryptocurrencies like Bitcoin, is poised to have a significant impact on technology entrepreneurship. Blockchain is a distributed ledger that allows secure and transparent transactions without the need for intermediaries.

Entrepreneurs can explore the potential of blockchain technology for creating decentralized applications, enabling secure peer-to-peer transactions, and enhancing supply chain management. Blockchain has the potential to revolutionize industries such as finance, logistics, and healthcare by providing increased transparency, efficiency, and trust.

However, challenges exist in terms of scalability, interoperability, and regulatory frameworks. Entrepreneurs must navigate these challenges and develop innovative solutions that harness the benefits of blockchain technology while mitigating associated risks.

Challenge 1: Access to Capital

Access to capital is a perennial challenge for technology entrepreneurs. Developing and scaling technology-based innovations often require substantial investments in research and development, talent acquisition, marketing, and infrastructure.

Entrepreneurs need to evaluate various funding options, such as angel investors, venture capitalists, crowdfunding, and government grants. They should also develop compelling business plans and communicate their value proposition effectively to attract potential investors.

Governments and financial institutions play a crucial role in creating an enabling ecosystem for technology entrepreneurship by providing favorable policies, incentives, and access to funding.

Challenge 2: Talent Acquisition and Retention

In today's technology-driven economy, attracting and retaining top talent is critical for the success of technology startups. Entrepreneurs face fierce competition in recruiting skilled professionals in areas such as software development, data science, and AI.

To overcome this challenge, entrepreneurs should focus on creating an inclusive and diverse work culture that values innovation, collaboration, and continuous learning. They can also forge partnerships with academic institutions to tap into talent pipelines and provide internships and mentorship programs.

Investing in employee training and development is essential for upskilling the workforce and retaining valuable talent. Additionally, flexible work arrangements and employee benefits can help attract and retain a diverse range of employees.

Challenge 3: Regulatory and Legal Hurdles

Technology entrepreneurship operates within a complex regulatory and legal landscape. Entrepreneurs need to navigate intellectual property rights, data protection regulations, and industry-specific compliance requirements.

Creating strategies to protect intellectual property is crucial for technology entrepreneurs. This includes obtaining appropriate patents, trademarks, and copyrights, as well as implementing robust data privacy and security measures.

Entrepreneurs should stay updated on changes in regulations and seek legal counsel to ensure compliance. Engaging with industry associations and advocacy groups can help entrepreneurs influence policy discussions and contribute to the development of supportive regulatory frameworks.

Challenge 4: Market Volatility and Competition

In the fast-paced world of technology entrepreneurship, market volatility and competition are constant challenges. Technological advancements and changing customer preferences can quickly render existing products or services obsolete.

Entrepreneurs must continually monitor market trends, anticipate disruptions, and adapt their business models accordingly. Adopting agile and iterative approaches to product development and maintaining a customer-centric focus are essential strategies for staying competitive.

Collaboration and strategic partnerships can help entrepreneurs leverage complementary resources and expand their market reach. Networking and staying connected with industry peers can provide valuable insights and opportunities for collaboration.

In conclusion, technology entrepreneurship is an exciting and dynamic field, driven by emerging trends and shaped by significant challenges. By understanding and embracing future trends such as AI, IoT, and blockchain, and by addressing challenges related to access to capital, talent acquisition, regulatory hurdles, and market volatility, entrepreneurs can position themselves for success in the ever-evolving technology-driven economy.

Technology, Innovation, and Economic Policy

Understanding the relationship between technology, innovation, and economic policy

Analyzing the role of government policies in promoting technological innovation

Government policies play a crucial role in promoting technological innovation, which is essential for economic growth and development. By creating an environment conducive to innovation, governments can encourage research and development activities, facilitate technology transfer, and provide support for entrepreneurs and startups. In this section, we will explore the various ways in which government policies can promote technological innovation.

Understanding the Importance of Technological Innovation

Before delving into the role of government policies, it is important to understand the significance of technological innovation. Technological innovation refers to the development and application of new or improved technologies that lead to the creation of new products, services, or processes. It is a key driver of economic growth, as it enhances productivity, increases competitiveness, and creates new market opportunities. Technological innovation also has the potential to address societal challenges, such as improving healthcare, combating climate change, and reducing poverty.

Analyzing the Policy Approaches

Governments can adopt different policy approaches to promote technological innovation. These approaches can include:

1. Funding research and development: Governments can allocate resources to support research and development activities in both public and private sectors. This can be done through grants, subsidies, and tax incentives, which encourage businesses and organizations to invest in innovation. By providing financial support, governments can reduce the risks associated with innovation and encourage experimentation.

2. Intellectual property rights protection: Intellectual property rights, such as patents, copyrights, and trademarks, play a crucial role in fostering innovation. Government policies can ensure the protection of these rights, providing inventors and innovators with the confidence to invest in research and development. A robust intellectual property framework promotes knowledge-sharing and encourages collaboration between academia, research institutions, and industry.

3. Creating innovation ecosystems: Governments can foster the development of innovation ecosystems by bringing together key stakeholders, including universities, research institutions, businesses, investors, and entrepreneurs. By facilitating collaboration and knowledge-sharing, governments can create an environment that supports the exchange of ideas, the development of new technologies, and the commercialization of innovations.

4. Strengthening education and skills development: To promote technological innovation, governments need to invest in education and skills development. By improving educational institutions and programs, governments can equip individuals with the necessary knowledge and skills to innovate. This includes STEM (Science, Technology, Engineering, and Mathematics) education, vocational training, and lifelong learning initiatives.

5. Promoting entrepreneurship: Governments can support entrepreneurship by creating policies that reduce barriers to entry, simplify administrative procedures, and provide access to finance and resources. Entrepreneurial activities often drive technological innovation as startups are more agile and willing to take risks. Government policies can assist startups by providing incubators, accelerators, and mentoring programs.

Case Studies

To better understand the role of government policies in promoting technological innovation, let's examine a few successful case studies:

1. Silicon Valley, United States: Silicon Valley is renowned for its thriving technological innovation ecosystem. The success of this region can be attributed in part to supportive government policies. The U.S. government has provided funding for research and development through agencies like the National Science Foundation and the Small Business Administration. Additionally, intellectual property rights protection has played a crucial role in attracting investment and fostering collaboration.

2. Israel's "Start-up Nation": Israel has gained international recognition for its vibrant startup ecosystem. Policies implemented by the Israeli government have played a significant role in fostering technological innovation. These policies include generous government funding for research and development, military service programs that cultivate technical expertise, and tax incentives for investors and entrepreneurs. The government has also facilitated collaborations between universities, research institutions, and businesses.

3. South Korea's Research and Development Investments: The South Korean government has made substantial investments in research and development, particularly in fields like information technology, biotechnology, and renewable energy. This proactive approach has helped South Korea become a leader in technological innovation. The government's emphasis on education and skills development has also contributed to a strong labor force with specialized expertise.

Challenges and Future Directions

While government policies play a crucial role in promoting technological innovation, several challenges need to be addressed:

1. Funding constraints: Governments often face constraints in allocating adequate resources for research and development. Limited funding can hinder the progress of technological innovation.

2. Regulatory environment: Excessive regulations can stifle innovation and make it challenging for startups and entrepreneurs to navigate the market. Governments need to strike a balance between ensuring public safety and promoting innovation-friendly regulations.

3. Market-oriented approach: Government policies should not substitute market mechanisms but rather complement them. An overreliance on government intervention may lead to inefficient outcomes.

In the future, government policies should evolve to address emerging trends and challenges in technological innovation. This includes supporting emerging technologies such as artificial intelligence, blockchain, and clean technologies. Governments should also focus on promoting inclusive and responsible innovation, ensuring that the benefits of technological advancements are equitably distributed and addressing any potential negative social or environmental impacts.

In conclusion, government policies play a vital role in promoting technological innovation. By providing funding, protecting intellectual property rights, creating innovation ecosystems, investing in education, and supporting entrepreneurship, governments can foster an environment that encourages research, development, and technological advancements. However, policymakers need to address challenges and adapt policies to ensure the continued promotion of technological innovation for sustainable economic growth and societal well-being.

Discussing the importance of research and development investments for technological advancements

Research and development (R&D) investments play a vital role in driving technological advancements. R&D involves the systematic investigation and experimentation aimed at acquiring new knowledge and developing innovative technologies. It plays a crucial role in promoting economic growth, fostering innovation, and addressing societal challenges. In this section, we will discuss the importance of R&D investments for technological advancements and explore the various ways in which they contribute to the development of new technologies.

The Role of R&D in Innovation

Innovation is the process of translating new knowledge into products, processes, or services that create value. R&D investments are a key driver of innovation, as they enable the generation of new ideas and the development of cutting-edge technologies. By investing in R&D, companies and organizations can explore new possibilities, improve existing technologies, and create entirely new ones.

The importance of R&D in fostering innovation can be seen in various sectors. For example, in the pharmaceutical industry, R&D investments are crucial for the development of new drugs and treatments. Pharmaceutical companies spend billions of dollars on R&D to conduct clinical trials, explore new therapeutic approaches, and discover novel compounds. These investments not only lead to the development of life-saving drugs but also contribute to advancements in medical science.

Similarly, in the technology sector, R&D investments drive innovation by enabling the creation of new products and services. Companies like Apple, Google, and Microsoft allocate significant resources to R&D to develop groundbreaking technologies such as smartphones, search engines, and operating systems. These innovations not only revolutionize industries but also have a significant impact on our daily lives.

Stimulating Economic Growth

R&D investments have a direct link to economic growth. They contribute to productivity gains, increased competitiveness, and job creation, all of which fuel economic development. When companies invest in R&D, they develop new technologies that improve efficiency, enhance productivity, and create new markets and industries.

One example of R&D investments driving economic growth is the development of renewable energy technologies. In recent years, there has been a significant increase in R&D investments in solar, wind, and other renewable energy sources. These investments have led to the development of more efficient and affordable renewable energy technologies, which in turn have stimulated the growth of the renewable energy sector and reduced dependence on fossil fuels. This transition not only creates jobs in the renewable energy industry but also promotes sustainable economic development.

R&D investments also contribute to economic growth by spurring innovation and entrepreneurship. Startups and small businesses often rely on R&D to develop innovative products and services that disrupt traditional industries. These innovations create new market opportunities, generate employment, and boost economic growth. Government policies that support R&D investments, such as tax incentives and grants, can further stimulate innovation and economic development.

Addressing Societal Challenges

R&D investments are instrumental in addressing societal challenges and finding solutions to complex problems. By allocating resources to R&D, governments, organizations, and researchers can develop technologies that tackle issues related to healthcare, climate change, food security, and more.

In the healthcare sector, R&D investments enable the development of new diagnostic tools, treatments, and therapies. For example, the significant investments in R&D during the COVID-19 pandemic led to the rapid development of vaccines in record time. Similarly, R&D investments in biotechnology have resulted in advancements in genetic engineering, personalized medicine, and regenerative therapies, offering new treatment options for various diseases and improving patient outcomes.

R&D investments also play a crucial role in addressing environmental challenges. For instance, investments in clean energy technologies, such as solar panels and electric vehicles, help reduce greenhouse gas emissions and mitigate climate change. Moreover, R&D in agriculture can lead to the development of sustainable farming practices, crop improvement, and efficient food production systems, contributing to global food security.

Challenges and Considerations

While R&D investments bring tremendous benefits, they also come with challenges and considerations. One significant challenge is the high cost associated with research and development. Developing new technologies often requires substantial financial resources, laboratory facilities, and highly skilled researchers. This cost can be a barrier, particularly for small businesses and developing economies.

Moreover, R&D investments are inherently risky. Not all research projects lead to successful outcomes or commercial viability. It requires a long-term perspective and a tolerance for failure, as breakthroughs often come after many iterations and setbacks. Balancing short-term financial considerations with the long-term potential of R&D investments can be a complex challenge for organizations and policymakers.

Additionally, intellectual property rights and patent protection play a crucial role in incentivizing R&D investments. Companies and researchers need assurance that they will be able to recoup their investments and benefit from the commercialization of their innovations. Developing effective mechanisms for intellectual property protection and balancing the need for openness and collaboration with the need for incentives is a critical consideration in fostering R&D investments.

Conclusion

R&D investments are crucial for technological advancements and have a profound impact on economic growth, innovation, and the resolution of societal challenges. By fostering innovation, stimulating economic growth, and addressing complex problems, R&D investments contribute to the betterment of society as a whole. However, it is essential to recognize the challenges associated with R&D investments and develop strategic approaches to overcome them. With appropriate policies, funding mechanisms, and collaboration between governments, organizations, and researchers, R&D investments can continue to drive technological advancements and shape a brighter future.

Examining the role of intellectual property rights in fostering innovation

Innovation is a key driver of economic growth and development. It fuels technological advancements, enhances productivity, and creates new opportunities for businesses and individuals. Intellectual property (IP) rights play a crucial role in fostering innovation by providing legal protection and incentives for inventors and creators. In this section, we will examine the importance of intellectual property rights in promoting innovation and explore the various types of IP protection available.

Understanding Intellectual Property Rights

Intellectual property refers to the creations of the human mind, such as inventions, literary and artistic works, symbols, names, images, and designs, that can be protected under various legal frameworks. Intellectual property rights grant exclusive rights to the creators or owners of these creations, enabling them to control and profit from their innovations. There are several types of intellectual property rights, including patents, copyrights, trademarks, and trade secrets.

Patents

Patents are one of the most widely recognized forms of intellectual property rights. A patent gives inventors exclusive rights to their inventions for a limited period, typically 20 years from the filing date. By granting inventors the right to prevent others from making, using, or selling their inventions without permission, patents incentivize innovation by providing a means for inventors to recoup their investments and reap the rewards of their ingenuity.

Patents play a vital role in fostering innovation in various industries. They encourage research and development (R&D) by providing inventors with a monopoly over their inventions, which creates a competitive advantage. This exclusive right stimulates inventors to disclose their inventions to the public, contributing to the spread of knowledge and further technological advancement. Patents also facilitate technology transfer and licensing agreements, enabling the diffusion of innovation from one entity to another.

However, the patent system is not without its challenges. The process of obtaining a patent can be complex, time-consuming, and expensive. Patent litigation and disputes can also arise, especially in highly competitive industries. Additionally, the patent system raises questions about the balance between rewarding inventors and ensuring that patented technologies are accessible for further development and public use.

Copyrights

Copyrights protect original works of authorship, such as literary, artistic, musical, and dramatic works, as well as software and databases. Unlike patents, which protect functional inventions, copyrights focus on the expression of ideas in a tangible form. Copyright protection grants creators exclusive rights to reproduce, distribute, and publicly display their works. The duration of copyright protection varies depending on the type of work and the jurisdiction, but it generally lasts for the life of the creator plus a certain number of years.

Copyrights play a vital role in fostering innovation in the creative industries. By providing creators with exclusive rights, copyrights encourage the creation of new works and allow creators to monetize their creations. This incentivizes artists, writers, musicians, and filmmakers to invest in their creative endeavors, ensuring a vibrant cultural and artistic landscape. Copyrights also promote the dissemination of knowledge and cultural heritage by granting creators the right to control the distribution and use of their works.

However, the advent of digital technologies and the internet has posed challenges to copyright protection. The ease of reproducing and distributing digital content has led to concerns about unauthorized copying and piracy. Striking a balance between protecting the rights of creators and promoting access to knowledge and culture in the digital age remains a major challenge for copyright laws and policies.

Trademarks

Trademarks are distinctive signs, such as words, logos, or symbols, that identify and distinguish goods or services of one entity from those of others. Trademark registration grants exclusive rights to use and protect the trademark, providing legal recourse against unauthorized use or infringement. Trademarks play a crucial role in building brand recognition and consumer trust, and they contribute to fair competition by preventing consumer confusion.

Trademarks provide incentives for businesses to invest in the development and marketing of innovative products and services. By creating a strong brand identity, trademarks enable businesses to differentiate themselves in the marketplace, attract customers, and establish a competitive edge. Trademarks also encourage businesses to maintain the quality and reputation associated with their brands, fostering consumer loyalty and trust.

However, the globalization of markets and the rise of e-commerce have presented new challenges for trademark protection. Online counterfeiting and infringement have become prevalent issues, requiring effective enforcement mechanisms and international cooperation to combat these threats. Additionally, the registration process for trademarks can be complex, and conflicts may arise when similar trademarks coexist in the marketplace.

Trade Secrets

Trade secrets are valuable and confidential business information that derives its economic value from remaining secret. Examples of trade secrets include manufacturing processes, formulas, customer lists, marketing strategies, and business plans. Unlike patents or copyrights, trade secrets rely on confidentiality rather than legal protection. As long as the information remains secret and the owner takes reasonable steps to maintain its secrecy, trade secrets can be protected indefinitely.

Trade secrets incentivize innovation by providing businesses with a competitive advantage. By keeping valuable information confidential, companies can gain a head start in the market, attract investors, and secure a unique position. Trade secrets also facilitate collaborations and partnerships, as businesses may be more willing to enter into agreements where sensitive information is protected as a trade secret.

However, trade secret protection can be challenging to enforce, especially in cases of misappropriation or unauthorized disclosure. Maintaining secrecy requires a robust and comprehensive set of security measures, including restricted access, non-disclosure agreements, and employee training. Companies must strike a balance between protecting trade secrets and sharing information for collaborative purposes.

The Role of Intellectual Property Rights in Fostering Innovation

Intellectual property rights play a crucial role in fostering innovation by providing legal protection, incentives, and rewards for creators and inventors. They encourage R&D investments, promote knowledge sharing, and facilitate technology transfer. Intellectual property protection enables inventors and creators to recoup their investments and fuel further innovation.

However, it is essential to strike a balance between protecting intellectual property and ensuring that knowledge and innovations are accessible for further development and public use. Overly restrictive intellectual property regimes can stifle competition, hinder creativity, and limit access to critical technologies. Therefore, policymakers must consider the broader societal implications of intellectual property rights and establish frameworks that encourage innovation while safeguarding the public interest.

Example Case Study: Intellectual Property Rights in the Pharmaceutical Industry

One prominent example of the role of intellectual property rights in fostering innovation is the pharmaceutical industry. Patents are crucial for pharmaceutical companies as they allow for the exclusive production and sale of drugs, protecting the significant investments made in drug development.

Pharmaceutical companies invest substantial resources in research, clinical trials, and regulatory approvals to bring new drugs to market. The high costs and risks involved in drug development necessitate a period of exclusivity to recoup investments and generate profits. Patents provide this exclusivity by granting pharmaceutical companies the right to exclude others from producing or selling their patented drugs without permission.

The protection provided by patents incentivizes pharmaceutical companies to invest in R&D activities, leading to the development of innovative drugs that save lives, alleviate suffering, and improve public health. Patent protection also encourages the sharing of scientific knowledge through the disclosure of inventions in patent applications, enabling other researchers to build upon existing discoveries.

However, the high prices of patented drugs and access issues in low-income countries raise questions about the balance between patent protection and affordable access to essential medicines. Striking a balance between rewarding innovation and ensuring access to life-saving drugs remains a significant challenge for policymakers, international organizations, and public health advocates.

Conclusion

Intellectual property rights, including patents, copyrights, trademarks, and trade secrets, play a critical role in fostering innovation. They provide legal protection, incentives, and rewards for inventors and creators, encouraging investment in R&D activities and the dissemination of knowledge. However, policymakers must strike a balance between protecting intellectual property and promoting access to innovations for the benefit of society as a whole.

Understanding the impact of competition policy on technology and innovation

Competition policy plays a crucial role in shaping the landscape of technology and innovation. It is a set of rules and regulations that aim to promote healthy competition in the market, prevent anti-competitive practices, and protect the interests of both consumers and businesses. In the context of technology and innovation, competition policy can have significant implications for fostering a conducive environment for technological advancements and promoting innovation. In this section, we will explore the impact of competition policy on technology and innovation, discussing its various dimensions and the challenges associated with its implementation.

Competition policy and market structure

Competition policy affects the structure of markets, which plays a critical role in influencing technology and innovation. A competitive market environment encourages firms to invest in research and development (R&D) activities and fosters innovation. When markets are competitive, firms are motivated to differentiate themselves from their competitors by developing new and improved products, services, or production processes. This leads to increased innovation and technological advancements.

On the other hand, in markets with limited competition or monopolistic tendencies, firms may have less incentive to innovate. A lack of competition reduces the pressure to invest in R&D, as firms can rely on their position in the market to maintain profits without technological advancements. Therefore, an effective competition policy that promotes competition and prevents the abuse of market power is crucial for driving innovation and technological progress.

Promoting competition in technology markets

Technology markets have unique characteristics that warrant specific consideration in competition policy. These markets are often characterized by network effects, high fixed costs, and intellectual property rights. Network effects occur when the value of a product or service increases as more people use it. This can create barriers to entry, as it becomes challenging for new entrants to compete with established firms that have already accumulated a significant user base.

Competition policy should address these challenges by ensuring that dominant firms do not exploit their market power to hinder competition and innovation. One approach is to promote interoperability standards, enabling different technologies to work together and preventing the creation of proprietary ecosystems that restrict competition. Additionally, competition authorities can scrutinize mergers and acquisitions to prevent the consolidation of market power.

Balancing competition and intellectual property rights

Intellectual property rights (IPRs), such as patents, copyrights, and trademarks, play a crucial role in incentivizing innovation by granting exclusive rights to inventors and creators. However, competition policy needs to strike a balance between encouraging innovation through IPRs and preventing the abuse of those rights to stifle competition.

One challenge is the strategic use of patents as a competitive weapon instead of promoting innovation. Patent thickets, where multiple patents cover various aspects of a technology, can create barriers to entry and hinder competition. Competition authorities can address this issue by assessing the validity and essentiality of patents, preventing the misuse of IPRs to exclude competitors from the market.

Another aspect is the consideration of standard-essential patents (SEPs) that are crucial for implementing industry standards. Licensing practices related to SEPs can impact competition and innovation, as firms may engage in anti-competitive behaviors, such as refusing to license their patents or setting excessively high licensing fees. Competition policy can ensure fair and reasonable licensing terms for SEPs to ensure that innovation is not hindered by anti-competitive practices.

Addressing anti-competitive behavior in technology markets

Competition policy must also address anti-competitive practices in technology markets. One common concern is the abuse of dominant market positions by technology giants. Dominant firms may engage in anti-competitive behaviors, such as predatory pricing, exclusive dealing, or leveraging their market power in one area to dominate another.

Competition authorities can take various measures to address these issues. They can require dominant firms to provide fair access to essential facilities or data, ensuring a level playing field for competitors. Additionally, competition policy can facilitate market entry by reducing barriers, such as complex regulatory frameworks or intellectual property hurdles, that may impede competition and innovation.

The role of competition policy in fostering collaboration

Collaboration and cooperation between firms can also drive innovation and technological progress. However, collaboration can have anti-competitive effects if it reduces competition or leads to collusion. Competition policy must strike a balance between promoting collaboration for innovation and preventing anti-competitive practices.

One approach is to establish guidelines and frameworks that facilitate collaboration while ensuring that it does not harm competition. Examples include technology-sharing agreements, joint R&D ventures, and collaborations for standard setting. Competition authorities can monitor and assess such collaborations to ensure that they do not result in anti-competitive outcomes.

Challenges in implementing effective competition policy

Implementing effective competition policy in the technology and innovation sector poses several challenges. One of the significant challenges is the fast-paced nature of technological advancements, which often outpaces the ability of regulators to keep up with market developments. Policymakers must have a deep understanding of technology-related markets and be able to anticipate potential competition concerns.

Another challenge is the global nature of technology markets. Technology firms operate across borders, and anti-competitive practices can have international implications. Coordination and cooperation among competition authorities globally are essential to address cross-border competition concerns effectively.

Furthermore, the dynamic and complex nature of technology markets often requires regulatory frameworks to be flexible and adaptable. Traditional approaches to competition policy may not adequately address emerging issues, such as platform markets, data-driven business models, or algorithmic decision-making. Policymakers need to continuously update competition frameworks to account for these evolving market dynamics.

Conclusion

Competition policy plays a vital role in shaping technology and innovation ecosystems. An effective competition policy promotes competition, prevents anti-competitive practices, and encourages innovation in technology markets. By ensuring a level playing field, addressing market concentration, and balancing intellectual property rights, competition policy can create an environment that fosters technological advancements. However, implementing effective competition policy in technology markets comes with challenges, such as keeping pace with rapid technological changes and addressing global competition concerns. Policymakers must continuously adapt and refine competition frameworks to effectively promote competition and innovation in the technology sector.

Investigating the role of government support programs in promoting innovation

Innovation plays a vital role in driving economic growth and competitiveness. Governments around the world recognize the importance of innovation and have implemented various support programs to foster innovation in their economies. These government support programs aim to encourage research and development (R&D) activities, facilitate technology transfer, and provide financial assistance to innovative ventures. In this section, we will investigate the role of government support programs in promoting innovation and their impact on economic development.

Understanding the importance of government support programs

Government support programs have become crucial in promoting innovation due to the high costs and risks associated with R&D activities. These programs offer financial, technical, and educational assistance to individuals, startups, and established businesses to encourage them to invest in innovative projects. By providing support in the form of grants, subsidies, tax incentives, or loans, governments aim to reduce the financial burden on innovators and create an enabling environment for innovation to thrive.

Moreover, government support programs help bridge the gap between academia and industry by promoting technology transfer and collaboration between research institutions and businesses. By fostering partnerships and knowledge exchange, these programs ensure that research outcomes are translated into practical applications and commercial products. This encourages the adoption of new technologies and facilitates their integration into various industries.

Types of government support programs

Government support programs for innovation can take various forms, depending on the specific goals and priorities of the government. Some common types of support programs include:

1. **Grants and subsidies:** These programs provide financial support to businesses and research institutions engaged in R&D activities. Grants can

cover a portion of the project costs, enabling businesses to invest in innovative projects that they may not have been able to finance independently. Subsidies, on the other hand, reduce the cost of inputs or resources required for innovation, making it more affordable for businesses to undertake research or adopt new technologies.

2. **Tax incentives:** Governments often use tax incentives as a means to promote innovation. These incentives can include tax credits for R&D expenditures, reduced tax rates for income generated from intellectual property, or tax exemptions for businesses operating in specific sectors or regions. By reducing the tax burden, these incentives encourage businesses to invest in R&D activities and drive innovation.

3. **Technology procurement:** Governments can play a significant role as early adopters of innovative technologies. By procuring goods and services that incorporate new technologies, governments create a market demand for innovative solutions, which encourages businesses to invest in research and development. This approach provides businesses with a stable customer base and can help validate and commercialize their innovative products.

4. **Incubation and acceleration programs:** These programs provide startups and entrepreneurs with essential resources, such as funding, mentoring, networking opportunities, and access to shared facilities. Incubators and accelerators facilitate the growth and development of innovative ventures by offering a supportive ecosystem and connecting them with potential investors and customers. Governments often establish and fund these programs to foster entrepreneurial activities and promote innovation-driven startup culture.

5. **Intellectual property support:** Government support programs also include initiatives aimed at protecting and managing intellectual property rights. These programs help businesses navigate the complex realm of patenting, licensing, and copyright, enabling them to safeguard their innovations and leverage their intellectual assets for commercialization and market competitiveness. By providing legal and technical assistance, governments foster an environment that encourages innovation and protects the interests of innovators.

Success stories of government support programs

Several countries have successfully implemented government support programs to foster innovation and drive economic growth. Let's explore a few success stories:

1. **Singapore's Research, Innovation and Enterprise (RIE) 2025 Plan:** Singapore has been consistently ranked among the top innovative countries globally. The RIE 2025 Plan, launched by the Singaporean government, aims to invest SGD 25 billion (approximately USD 18.7 billion) over five years in R&D and innovation. The plan focuses on developing key technological capabilities, fostering collaboration between public and private sectors, and promoting entrepreneurship and innovation-driven enterprises.

2. **Israel's Office of the Chief Scientist (OCS):** The OCS, under the Israeli Ministry of Economy, has successfully promoted innovation through various support programs. The OCS offers financial assistance, grants, and tax incentives to startups and businesses engaged in R&D activities. It also provides support for technology transfer, international collaboration, and the creation of incubators and research centers. Israel's thriving startup ecosystem and technological advancements can be attributed, in part, to these government support programs.

3. **Finland's Tekes – The Finnish Funding Agency for Innovation:** Tekes is a governmental funding agency that supports innovation and R&D activities in Finland. Tekes provides funding and expert guidance to businesses of all sizes, from startups to large enterprises. It focuses on promoting disruptive innovations, fostering strategic partnerships, and supporting the development of emerging technologies. The success of Finnish companies, such as Nokia and Rovio (creator of Angry Birds), can be attributed to Tekes' support.

Key challenges and considerations

While government support programs play a crucial role in promoting innovation, there are certain challenges and considerations that need to be addressed:

- **Effectiveness and evaluation:** It is essential to continuously evaluate the effectiveness of government support programs to ensure optimal allocation of resources. Programs should be periodically assessed based on their impact on innovation outcomes, job creation, economic growth, and sustainability. Governments must adapt their programs based on the lessons learned and evolving innovation landscape.

- **Risks and accountability:** Government support programs involve risks, particularly financial risks associated with failed projects or misuse of funds. It is crucial for governments to have transparent and accountable processes in place to mitigate these risks and ensure that funds are allocated to the most promising and economically viable projects. Proper oversight and accountability mechanisms are necessary to maintain trust and confidence in the programs.

- **Coordination and collaboration:** Governments need to coordinate their support programs with other stakeholders, such as academia, industry associations, and international organizations. Collaboration between different actors in the innovation ecosystem can leverage resources and expertise, foster knowledge sharing, and avoid duplication of efforts. Alignment with national innovation strategies and goals is also essential for the success of these programs.

- **Inclusive innovation:** Governments should aim for inclusive innovation, ensuring that support programs are accessible to all segments of society, including underrepresented groups and regions. It is crucial to promote diversity in innovation and create equal opportunities for participation. Measures should be in place to address the digital divide, gender disparities, and geographical imbalances in innovation activities.

Conclusion

Government support programs play a vital role in promoting innovation by providing financial, technical, and educational assistance to businesses and research institutions. These programs reduce the barriers to entry for innovation and create an enabling environment for research, development, and technology transfer. Through successful support programs, countries have witnessed the emergence of thriving startup ecosystems, technological advancements, and economic growth. However, challenges such as program evaluation, risk management, coordination, and inclusivity need to be addressed to ensure the long-term success and effectiveness of these programs. Governments must continuously learn from best practices, adapt their programs, and foster collaboration to drive innovation, economic development, and sustainable growth.

Analyzing the challenges and opportunities of open innovation approaches

Open innovation is an approach that emphasizes collaboration and the sharing of ideas, knowledge, and resources among different entities, such as companies, researchers, and even individuals. It is based on the belief that innovation can come from a wide range of sources, both inside and outside the organization. In this section, we will explore the challenges and opportunities associated with open innovation approaches.

Challenges of open innovation

While open innovation offers numerous benefits, there are also challenges that need to be addressed for its successful implementation. Some of the key challenges include:

Intellectual property (IP) concerns: One of the main concerns in open innovation is the protection of intellectual property. When collaborating with external parties, there is a risk of knowledge leakage and unauthorized use of proprietary information. This poses a challenge for companies who want to engage in open innovation while safeguarding their IP rights.

Trust and collaboration: Open innovation requires a high degree of trust and collaboration among various stakeholders. Establishing trust can be challenging, especially when dealing with partners from different cultural backgrounds or industries. Moreover, collaboration among competitors may raise concerns about anti-competitive practices and the sharing of trade secrets.

Managing knowledge flow: In open innovation, managing the flow of knowledge becomes crucial. It is essential to strike a balance between disclosing enough information to engage potential collaborators and protecting sensitive information. Moreover, managing the transfer of knowledge in a structured and efficient manner can be challenging, particularly when dealing with a large number of partners.

Cultural and organizational barriers: Open innovation may face resistance within organizations due to a culture that values secrecy and a reluctance to let external parties participate in the innovation process. Overcoming these cultural barriers and fostering a mindset that embraces collaboration and external contributions requires strong leadership and a supportive organizational culture.

Lack of alignment: In open innovation, different parties may have different objectives, priorities, and timelines. This lack of alignment can create challenges in terms of managing expectations, setting common goals, and ensuring effective coordination among partners.

Managing complexity: Open innovation often involves multiple partners with diverse expertise, resources, and capabilities. Managing the complexity arising from these diverse collaborations can be a significant challenge, particularly when dealing with large-scale and long-term projects.

Opportunities of open innovation

Despite the challenges, open innovation offers several opportunities that can enhance the innovation ecosystem and drive economic growth. Some of the key opportunities include:

Access to external knowledge and expertise: Open innovation allows organizations to tap into a vast pool of external knowledge and expertise. Collaborating with diverse partners, such as research institutions, startups, and customers, can bring new perspectives and insights that may not exist within the organization.

Accelerated innovation and time-to-market: By leveraging the capabilities of external partners, companies can accelerate the innovation process and reduce time-to-market. Open innovation enables access to ready-made solutions, technologies, and resources, allowing organizations to develop products and services more rapidly and efficiently.

Reduced costs and shared risks: Open innovation enables organizations to share the costs and risks associated with innovation. Collaborative efforts can help distribute the financial burden, making innovation more accessible to smaller organizations or those with limited resources. Additionally, pooling resources and capabilities can help mitigate the risks associated with innovation projects.

Expanded market reach and customer insights: Open innovation provides opportunities to access new markets and customer segments. Collaboration with partners who have different market knowledge and distribution channels can help organizations expand their reach and access new customer insights. This can lead to the development of more customer-centric products and services.

Fostering a culture of innovation: Open innovation promotes a culture of continuous learning, collaboration, and creativity. It encourages organizations to break down silos and embrace diverse perspectives and ideas. This can result in increased employee engagement and motivation, as well as a more innovative and adaptable organizational culture.

Increased competitiveness and industry leadership: Engaging in open innovation can enhance the competitiveness of organizations by leveraging external knowledge and resources. By staying connected to the latest developments and trends in the industry, organizations can position themselves as leaders and stay ahead of the competition.

Leveraging network effects: Open innovation allows organizations to tap into the network effects created by collaborating with various partners. A larger network of collaborators can lead to increased knowledge sharing, cross-pollination of ideas, and faster diffusion of innovation. This can create a virtuous cycle of innovation, where the more partners collaborate, the greater the benefits for all participants.

Case study: Open innovation in the pharmaceutical industry

One of the industries that have embraced open innovation is the pharmaceutical industry. Traditionally, pharmaceutical companies have operated in a closed innovation system, conducting research and development in-house and keeping the knowledge and expertise within the organization. However, with the increasing complexity and cost of drug development, many pharmaceutical companies have recognized the need to collaborate with external partners to drive innovation.

For example, in 2014, GlaxoSmithKline (GSK) launched the Open Lab initiative in Tres Cantos, Spain. The Open Lab is a collaborative research facility that allows scientists from GSK and external partners, such as academic institutions and nonprofits, to work together on neglected tropical diseases. By opening up its research facility and leveraging the expertise of external partners, GSK aims to accelerate the discovery of new treatments for diseases that disproportionately affect the developing world.

This open innovation approach has brought several benefits to GSK. By collaborating with external partners, GSK has gained access to new research capabilities, expertise, and fresh perspectives. The collaboration has not only accelerated the drug discovery process but has also reduced costs and risks associated with in-house research and development. Moreover, the Open Lab initiative has enhanced GSK's reputation as an innovative and socially responsible company.

However, implementing open innovation in the pharmaceutical industry comes with its own set of challenges. In addition to the intellectual property concerns, ensuring compliance with regulations, maintaining confidentiality, and managing complex collaborations are critical challenges that pharmaceutical companies must address. Nonetheless, the opportunities provided by open innovation, such as access to diverse knowledge, sharing of resources and risks, and accelerated drug development, make it a compelling approach in the industry.

Conclusion

Open innovation approaches offer both challenges and opportunities for organizations seeking to drive innovation and economic growth. By addressing the challenges associated with intellectual property concerns, trust and collaboration, knowledge flow, cultural barriers, lack of alignment, and managing complexity, organizations can unlock the full potential of open innovation. Leveraging the opportunities of accessing external knowledge and expertise, accelerating innovation and time-to-market, reducing costs and sharing risks, expanding market reach and customer insights, fostering a culture of innovation, and increasing competitiveness, organizations can harness the power of open innovation to stay competitive in an increasingly dynamic and interconnected world. The case study on open innovation in the pharmaceutical industry exemplifies the benefits and challenges associated with implementing open innovation approaches, highlighting the transformative impact such approaches can have on various industries.

Discussing the Importance of Technology Transfer and Commercialization

Technology transfer and commercialization play a crucial role in the successful integration of technological innovations into the marketplace. It involves the process of transferring knowledge, technologies, and intellectual property from research institutions or organizations to the private sector for commercial application. This section will discuss the importance of technology transfer and commercialization in driving economic growth, promoting innovation, and addressing societal challenges.

Driving Economic Growth

Technology transfer and commercialization contribute significantly to economic growth by stimulating investment, creating job opportunities, and generating wealth. When universities, research institutions, or technology companies engage in technology transfer activities, they not only monetize their intellectual property but also promote entrepreneurship and startup creation.

One of the primary benefits of technology transfer is the potential for generating revenue through licensing agreements and royalties. By licensing their technologies to companies, research institutions can generate income that can be reinvested in further research and development activities. This creates a cycle of innovation and economic growth.

Furthermore, technology transfer leads to the creation of spin-off companies and startups, which can boost employment opportunities and foster regional economic development. These start-ups often build upon advanced technologies developed in research institutions, commercialize them, and create new products and services that stimulate economic activity.

Promoting Innovation

Technology transfer and commercialization play a crucial role in promoting innovation by bridging the gap between academic research and industry. The translation of research findings into commercial products or processes enables the practical application of scientific knowledge. This application-driven research not only leads to technological advancements but also encourages further scientific inquiry.

Collaboration between industry and academia enables researchers to gain insights into practical challenges and industry needs. This interaction fosters a culture of innovation, encourages multidisciplinary collaboration, and provides real-world validation for research findings. Through technology transfer, industry partners can also provide funding, resources, and expertise to support research projects, accelerating the development of innovative solutions.

Furthermore, technology transfer facilitates the exchange of knowledge and expertise between different sectors. It enables cross-pollination of ideas, encourages knowledge spillovers, and creates opportunities for collaborative research and development. Through technology transfer, research findings from one field can be applied to solve problems in another, leading to breakthrough innovations.

Addressing Societal Challenges

Technology transfer and commercialization play a vital role in addressing societal challenges, such as healthcare, agriculture, and environmental sustainability. By facilitating the translation and commercialization of research findings, technology transfer enables the development of innovative solutions to these challenges.

In the healthcare sector, technology transfer has led to the development of new drugs, medical devices, and treatment methods. It enables the transfer of knowledge from academic research to pharmaceutical and medical device companies, accelerating the development and availability of life-saving technologies.

In agriculture, technology transfer promotes the adoption of sustainable farming practices, crop improvement techniques, and efficient resource management. By transferring knowledge and technologies to farmers and agricultural companies, technology transfer contributes to improving food security, reducing environmental impact, and increasing productivity.

Additionally, technology transfer plays a critical role in environmental sustainability. It enables the dissemination and adoption of clean technologies, renewable energy solutions, and waste management systems. By commercializing these technologies, technology transfer helps address climate change, reduce pollution, and promote a sustainable future.

Challenges and Strategies

While technology transfer and commercialization offer significant opportunities, there are also challenges that need to be addressed to ensure successful outcomes. Some of the key challenges include:

- Intellectual property management and protection
- Identification of potential industry partners
- Access to funding for technology development and commercialization
- Commercialization barriers, such as regulatory approvals and market acceptance
- Limited resources and expertise in technology transfer offices

To overcome these challenges and maximize the benefits of technology transfer and commercialization, several strategies can be employed:

- Strengthening the collaboration between industry and academia through joint research projects, consortia, and partnerships
- Enhancing intellectual property management and protection mechanisms to incentivize technology transfer
- Providing support and resources for technology transfer offices to streamline the commercialization process
- Facilitating access to funding through grants, venture capital, and government initiatives

- Promoting entrepreneurship and startup creation through incubators, accelerators, and mentorship programs

- Fostering a culture of innovation and knowledge-sharing within research institutions and companies

- Developing policies and regulations that support technology transfer and address commercialization barriers

- Promoting international collaborations and knowledge exchange to leverage global expertise and market opportunities

Successful case studies of technology transfer and commercialization can serve as valuable examples and sources of inspiration for researchers, entrepreneurs, and policymakers alike. By learning from these success stories, stakeholders can identify effective strategies and best practices to encourage and facilitate technology transfer activities.

Conclusion

Technology transfer and commercialization are imperative for driving economic growth, promoting innovation, and addressing societal challenges. Through the transfer of knowledge, technologies, and intellectual property, research findings can be transformed into practical applications that benefit society. To fully harness the potential of technology transfer and commercialization, it is essential to overcome challenges and implement effective strategies. By fostering collaboration, enhancing intellectual property protection, and providing support for technology transfer activities, we can create a thriving ecosystem that facilitates the successful commercialization of innovations.

Examining the Role of International Collaborations in Fostering Technological Innovation

In today's interconnected world, international collaborations play a crucial role in fostering technological innovation. As global challenges become increasingly complex, no single country can tackle them alone. Collaborative efforts, involving the exchange of knowledge, resources, and expertise, are essential for driving technological advancements and addressing global issues. This section will discuss the importance of international collaborations in fostering technological innovation and provide insights into successful examples and strategies.

Understanding the Need for International Collaborations

Technological innovation is a result of the collective knowledge and efforts of individuals, organizations, and countries worldwide. When countries collaborate, they gain access to a broader pool of talent, expertise, and resources, which accelerates the pace of innovation. International collaborations provide opportunities for scientists, engineers, and entrepreneurs to collaborate on research, development, and commercialization projects.

Furthermore, international collaborations allow for the sharing of best practices, experiences, and lessons learned. This cross-pollination of ideas and approaches leads to the creation of new solutions and the identification of innovative opportunities.

Benefits of International Collaborations in Technological Innovation

International collaborations bring about several benefits in the field of technological innovation:

1. **Access to diverse perspectives and expertise:** Collaborating with experts from different countries and cultures brings in fresh perspectives and diverse skill sets. This diversity can lead to breakthrough ideas and novel approaches to problem-solving, fostering innovation.

2. **Leveraging complementary strengths:** Countries often possess different strengths and resources in terms of research capabilities, infrastructure, funding, and market access. By collaborating, countries can leverage each other's strengths, filling gaps and enhancing their collective ability to innovate.

3. **Shared costs and reduced risks:** Technological innovation often requires significant investment in research, development, and commercialization. International collaborations allow countries to share costs and risks, making innovation more feasible and affordable for all parties involved.

4. **Accelerated innovation cycles:** Collaborations enable researchers and innovators to work together more efficiently, leveraging each other's progress and building on existing work. This leads to accelerated innovation cycles, enabling quicker development and adoption of new technologies.

5. **Access to international markets:** International collaborations facilitate access to new markets by establishing networks and partnerships across

borders. This not only enhances the commercialization of innovative technologies but also helps economies thrive through international trade and technology transfer.

Successful Examples of International Collaborations

Numerous successful examples demonstrate the power of international collaborations in fostering technological innovation. Here are a few prominent cases:

1. **The Human Genome Project (HGP)**: The HGP, launched in 1990, involved scientific collaborations between research institutions in the United States, the United Kingdom, and other countries. This ambitious project aimed to sequence and map the entire human genome. The collaborative efforts led to groundbreaking discoveries and advancements in genomics, enabling new insights into human health and disease.

2. **The International Space Station (ISS)**: The ISS is a multinational project involving space agencies from the United States, Russia, Europe, Japan, and Canada. This collaborative effort serves as a state-of-the-art research laboratory in space, fostering technological innovations that benefit multiple fields, including materials science, life sciences, and Earth observation.

3. **The European Organization for Nuclear Research (CERN)**: CERN, based in Switzerland, is a prime example of international collaboration in the field of particle physics. It brings together scientists and engineers from around the world to conduct cutting-edge research and develop advanced technologies. CERN's discoveries, such as the Higgs boson, have paved the way for new technologies and applications.

These successful examples highlight the transformative impact of international collaborations on technological innovation, demonstrating what can be achieved through shared efforts and resources.

Strategies for Facilitating Effective International Collaborations

To foster technological innovation through international collaborations, it is essential to adopt effective strategies. Here are some key strategies:

1. **Establishing strong networks and partnerships**: Building networks and partnerships between governments, research institutions, and industry

players facilitates collaboration. This can be done through bilateral agreements, research consortiums, and joint funding programs. Open communication channels and regular knowledge exchange platforms should be established to nurture cooperation.

2. **Promoting interdisciplinary collaboration:** Technological innovation often requires expertise from multiple disciplines. Encouraging interdisciplinary collaborations, spanning fields like science, engineering, social sciences, and business, helps generate holistic and innovative solutions to complex challenges.

3. **Supporting capacity building and technology transfer:** International collaborations should focus not only on research collaboration but also on capacity building and technology transfer. Providing training programs, workshops, and internships enables individuals and organizations to acquire new skills and knowledge that can be applied to drive domestic innovation.

4. **Creating enabling policy environments:** Governments play a crucial role in creating policies and regulations that facilitate international collaborations. Simplifying visa processes, harmonizing intellectual property rights protection, and providing financial incentives for collaborative projects are examples of policy measures that can promote cross-border collaboration.

It is important to note that while international collaborations offer significant benefits, they also present challenges. Differences in language, cultural norms, intellectual property rights, and funding mechanisms may create obstacles. Addressing these challenges requires effective communication, mutual respect, and a shared vision among the collaborating parties.

Ethical Considerations in International Collaborations

International collaborations in technological innovation raise ethical considerations that need to be carefully addressed. These include:

1. **Ownership and control:** Collaborations involve the sharing of knowledge, resources, and intellectual property. It is crucial to establish fair and transparent mechanisms for sharing benefits and ensuring that the interests of all parties involved are protected.

2. **Equitable access to benefits:** International collaborations should consider the equitable distribution of benefits arising from technological innovation.

This includes ensuring that the fruits of collaboration are accessible to all stakeholders, including researchers, innovators, and the wider society.

3. **Respect for cultural and societal norms:** Collaborations should maintain respect for cultural and societal norms, ensuring that the research and innovation activities align with the values and aspirations of all participating countries.

4. **Responsibility towards global challenges:** International collaborations should prioritize addressing global challenges, such as climate change, poverty, and inequality. Collaborative efforts must be guided by the principles of sustainability, social responsibility, and inclusivity.

By addressing these ethical considerations, international collaborations can foster technological innovation that not only benefits participating countries but also contributes to the broader global society.

Conclusion

In conclusion, international collaborations play a pivotal role in fostering technological innovation. By leveraging diverse perspectives, sharing resources, and promoting knowledge exchange, collaborations accelerate the pace of technological advancements. The benefits include access to expertise, shared costs and risks, accelerated innovation cycles, and access to international markets. Successful examples, such as the Human Genome Project, the International Space Station, and CERN, highlight the transformative impact of international collaborations. To facilitate effective collaborations, strategies like establishing strong networks, promoting interdisciplinary collaboration, and supporting capacity building should be adopted. However, ethical considerations, such as ownership and control, equitable access to benefits, and respect for cultural norms, must be carefully addressed. By embracing international collaborations while addressing the associated challenges, countries can collectively drive technological innovation for the betterment of society and the world at large.

Understanding the ethical implications of economic policies related to technology and innovation

In this section, we will explore the ethical implications of economic policies that are closely tied to technology and innovation. As technological advancements continue to shape our economies and societies, it is crucial to consider the ethical

UNDERSTANDING THE RELATIONSHIP BETWEEN TECHNOLOGY, INNOVATION, AND ECONOMIC POLICY

dimensions of the policies that govern them. We will discuss the importance of ethical considerations in economic decision-making, analyze specific ethical concerns related to technology and innovation, and propose strategies for addressing these challenges.

The Importance of Ethical Considerations

Ethics plays a fundamental role in guiding economic policies related to technology and innovation. Economic decisions have wide-ranging implications for individuals, communities, and the environment. Therefore, it is crucial to ensure that these decisions prioritize the well-being of all stakeholders and promote fairness, justice, and sustainability.

Ethics provides a framework for assessing the impact of economic policies on various dimensions, such as human rights, social equity, environmental protection, and privacy. By considering ethical principles, policymakers can make informed decisions that align with societal values and promote the common good.

Ethical Concerns in Technology and Innovation

1. Privacy and Data Protection: With the advent of digital technologies, the collection, storage, and analysis of personal data have become widespread. Economic policies must address the ethical concerns surrounding privacy and data protection. Individuals have the right to control their personal information and be protected from unauthorized access or misuse. Policymakers should establish robust regulations that ensure the responsible use of data and safeguard individuals' privacy.

2. Equity and Accessibility: Economic policies related to technology and innovation should aim to bridge the digital divide and promote equal opportunities for all. Access to technology and its benefits should not be limited to certain groups or regions. Policymakers should work towards ensuring equitable access to technology, digital infrastructure, and digital literacy training. This will help prevent exacerbating existing inequalities and ensure that the benefits of technological advancements are shared by all.

3. Ethical Use of Artificial Intelligence (AI) and Automation: The integration of AI and automation in various industries raises ethical concerns. Economic policies should address the potential risks associated with the use of these technologies, such as job displacement, algorithmic bias, and the concentration of power. Policymakers should establish guidelines and regulations to promote the ethical development and

deployment of AI and automation, ensuring that they serve human interests and uphold societal values.

4. Environmental Sustainability: Economic policies should also consider the environmental impact of technological innovations. The production and use of technology can contribute to resource depletion, pollution, and climate change. Policies should encourage the development and adoption of environmentally sustainable technologies, promote responsible resource management, and incentivize businesses to minimize their carbon footprint.

5. Responsible Innovation: Ethical considerations should be integrated into the innovation process itself. Economic policies should encourage responsible innovation that takes into account the potential risks and unintended consequences of new technologies. Policymakers should promote research ethics, encourage transparency in the development and testing of technologies, and establish mechanisms for ongoing assessment and accountability.

Addressing Ethical Concerns

Addressing the ethical implications of economic policies related to technology and innovation requires a comprehensive approach. Here are some strategies to consider:

1. Multi-stakeholder Engagement: Policymakers should involve various stakeholders, including industry experts, civil society organizations, academia, and the public, in the policymaking process. This ensures diverse perspectives are considered and allows for collective decision-making that promotes ethical considerations.

2. Ethical Impact Assessments: Economic policies should undergo ethical impact assessments to evaluate their potential consequences on different ethical dimensions. These assessments should be carried out in a transparent and participatory manner, considering both short-term and long-term impacts.

3. Robust Regulations: Policymakers should establish and enforce robust regulations that address specific ethical concerns related to technology and innovation. These regulations should aim to protect privacy, ensure fair competition, prevent algorithmic bias, and promote environmental sustainability.

4. Education and Awareness: Promoting ethical technology and innovation requires raising awareness and fostering ethical literacy. Policymakers should invest in educational programs and initiatives that promote ethical decision-making, digital literacy, and responsible use of technology.

5. International Cooperation: Ethical implications of economic policies related to technology and innovation are not limited to national boundaries. Policymakers

should collaborate internationally to develop common ethical standards, share best practices, and address global challenges collectively.

It is essential for economic policies to consider the ethical implications of technology and innovation to ensure a just, inclusive, and sustainable future. By integrating ethical considerations into policymaking, we can harness the power of technology for the benefit of all while minimizing potential harms.

Assessing the effectiveness of different policy approaches in promoting technology and innovation

In order to foster technology and innovation, governments around the world implement various policy approaches to create an environment conducive to technological advancements. In this section, we will assess the effectiveness of different policy approaches in promoting technology and innovation.

Policy Approach 1: Research and Development (R&D) Investments

One important policy approach is the investment in research and development (R&D). Governments allocate funds to support scientific research and technological innovations in both public and private sectors. These investments contribute to the development of cutting-edge technologies and drive economic growth.

R&D investments serve multiple purposes. Firstly, they support the discovery and creation of new knowledge, which forms the foundation for technological innovation. Research funding enables scientists and engineers to explore new ideas, conduct experiments, and develop prototypes. These efforts often lead to breakthroughs and advancements across various sectors, including healthcare, energy, and information technology.

Moreover, R&D investments promote collaboration between academia, industry, and government. By establishing partnerships and fostering knowledge exchange, these initiatives facilitate the transfer of ideas and technologies from research institutions to the marketplace. This collaboration helps overcome barriers to innovation and accelerates the adoption of new technologies in commercial settings.

Additionally, government-led R&D initiatives help address market failures related to innovation. Sometimes, private organizations may not invest in long-term and high-risk research projects due to uncertainty or lack of financial incentives. In such cases, government support becomes crucial to bridge the innovation gap. Public funding for R&D projects can stimulate private sector participation and encourage technology-driven entrepreneurship.

To assess the effectiveness of R&D investments, policymakers evaluate the impact of these initiatives on technological advancements, economic growth, and societal well-being. Measuring the number of patents, publications, and citations is one way to gauge the success of R&D investments. Furthermore, policymakers examine the commercialization of research outcomes, such as the number of new products, services, and companies launched as a result of research funding.

Challenges associated with R&D investments include ensuring the effective allocation of resources and minimizing the risk of duplication. Policymakers must also consider the ethical implications of R&D investments, such as the responsible and equitable use of emerging technologies.

Policy Approach 2: Intellectual Property Rights (IPR)

Intellectual Property Rights (IPR) protection is another policy approach aimed at promoting technology and innovation. IPR grants legal rights to creators and innovators, protecting their inventions, designs, and discoveries from unauthorized use or reproduction. By providing exclusive rights for a limited period, IPR incentivizes innovation and facilitates the transfer of technology.

One key aspect of IPR is patents, which allow inventors to exclude others from using, making, or selling their inventions for a specific period. Patents encourage inventors to disclose their innovations, which contributes to the overall body of knowledge and facilitates future technological advancements. Furthermore, patent holders can license their technology to other organizations, fostering knowledge diffusion and technology transfer.

Another form of IPR is copyright, which protects original literary, artistic, and creative works. Copyright law encourages authors and creators to produce new works by granting them exclusive rights over their creations. This protection incentivizes innovation in fields such as music, literature, software, and film, fostering a thriving creative industry.

Trademark protection is yet another form of IPR, which safeguards distinctive signs or symbols used by companies to distinguish their products or services. Trademarks ensure brand recognition and consumer trust, incentivizing companies to invest in product quality and innovation.

Efforts to assess the effectiveness of IPR policies include measuring the number of patents, copyrights, and trademarks registered. Economists also analyze the impact of IPR on investment, entrepreneurship, and knowledge diffusion.

Additionally, policymakers consider the balance between IPR protection and the accessibility of knowledge for further research and development.

However, some challenges exist in implementing IPR policies. Striking the right balance between providing incentives for innovation and ensuring fair access to knowledge is crucial. Policymakers need to consider the potential for IPR abuse and address concerns related to monopoly power and market concentration.

Policy Approach 3: Competition Policy

Competition policy plays a vital role in promoting technology and innovation by fostering vibrant market competition. By preventing anti-competitive practices and ensuring a level playing field, competition policy encourages companies to invest in research, development, and innovation.

One aspect of competition policy is the regulation of mergers and acquisitions. Antitrust authorities assess the potential impact of mergers on market competition, innovation, and consumer welfare. By scrutinizing these transactions, policymakers aim to prevent the formation of monopolies or dominant market positions that could stifle innovation or limit consumer choice.

Moreover, competition policy includes measures to curb anti-competitive practices such as collusion, price-fixing, and abuse of market dominance. These practices harm innovation by reducing incentives for companies to invest in research and development. Policymakers enforce strict penalties and fines for such anti-competitive behavior, creating a deterrent effect and ensuring fair competition.

To assess the effectiveness of competition policies, authorities monitor market dynamics, consumer prices, and innovation activities. They also evaluate the competitive landscape, market entry barriers, and the level of market concentration. Additionally, economists analyze the relationship between competition and innovation indicators, such as patents, R&D expenditures, and new product introductions.

However, implementing effective competition policies requires a comprehensive understanding of market dynamics and close collaboration between regulatory bodies and industry stakeholders. Policymakers must strike a balance between encouraging competition and ensuring market stability.

Policy Approach 4: Open Innovation

Open innovation is an approach that emphasizes collaboration and knowledge sharing among different organizations to foster technology and innovation. Unlike traditional closed innovation, where companies develop innovations internally, open innovation encourages the exchange of ideas, technologies, and expertise with external partners.

Open innovation promotes the idea that valuable knowledge exists both inside and outside an organization. By partnering with external entities, such as universities, research institutions, startups, and customers, companies can access a wider pool of expertise and resources. This collaboration enhances problem-solving capabilities, accelerates the pace of innovation, and reduces the risks and costs associated with internal R&D.

Policymakers support open innovation through policies that encourage knowledge transfer, facilitate technology licensing, and promote collaboration between public and private sectors. They establish innovation hubs, incubators, and accelerators that provide a conducive environment for knowledge exchange and support technology startups. Governments also incentivize organizations to share their intellectual property, ensuring that knowledge flows between academia, industry, and society.

Assessing the effectiveness of open innovation policies involves evaluating the quality and quantity of collaborative projects, analyzing the impact of knowledge sharing on innovation outputs, and measuring the success of technology transfer initiatives. Policymakers also consider the role of open innovation in fostering inclusive and sustainable development, promoting the involvement of diverse stakeholders and addressing societal challenges.

However, implementing open innovation policies requires addressing intellectual property rights, establishing trust and cooperation among participants, and overcoming barriers to collaboration. Policymakers must create mechanisms that facilitate the dissemination of knowledge while protecting the interests of participating organizations.

Conclusion

Assessing the effectiveness of different policy approaches in promoting technology and innovation is essential to guide policymakers in shaping effective strategies. R&D investments, intellectual property rights, competition policy, and open innovation are among the key policy approaches that influence technological advancements and foster innovation-driven economies.

By evaluating the impact of these policies on technological progress, economic growth, and societal well-being, policymakers can refine their approaches and address challenges associated with technology and innovation. Continual assessment and adaptation of policy approaches are crucial in a rapidly changing technological landscape.

In the next section, we will explore the impact of technology on employment and inequality, providing insights into the challenges and opportunities in creating inclusive and equitable technology-driven economies.

Technology, Employment, and Inequality

Examining the impact of technology on employment

Analyzing the displacement effect of technology on certain job sectors

In this section, we will explore the displacement effect of technology on certain job sectors. As technology continues to advance at an unprecedented pace, it is inevitable that some jobs will be replaced by automation, artificial intelligence, and other technological innovations.

Understanding the displacement effect

The displacement effect refers to the phenomenon in which technology replaces human workers in specific job sectors. This occurs when tasks previously performed by humans are automated or taken over by machines or software. The displacement effect can be seen across a wide range of industries, from manufacturing to customer service to transportation.

The displacement effect is driven by various factors. Firstly, technological advancements enable machines and software to perform tasks more efficiently and accurately than humans. This leads to increased productivity and cost savings for businesses. Secondly, the cost of implementing technology continues to decrease, making it more accessible and financially viable for organizations to adopt automation. Lastly, the demand for efficiency and competitiveness in the global market incentivizes companies to embrace technology to streamline their operations.

Implications for specific job sectors

While the displacement effect affects job sectors differently, some sectors are particularly vulnerable to widespread automation. Let's explore a few examples:

- **Manufacturing:** Automation has significantly reshaped the manufacturing industry, with robots replacing human labor in tasks such as assembly, packaging, and quality control. This has resulted in increased efficiency and lower production costs for companies, but it has also led to job losses in traditional manufacturing roles.
- **Retail:** The rise of e-commerce and advancements in technology have disrupted the retail industry. Self-checkout machines and online shopping platforms have reduced the need for human cashiers and sales associates. While new job opportunities have emerged in e-commerce logistics and customer support, the overall impact is a net loss of traditional retail jobs.
- **Transportation:** The advent of self-driving vehicles and delivery drones poses a significant threat to jobs in the transportation sector. Long-haul trucking, taxi services, and even delivery drivers are at risk of displacement. While these technologies offer potential benefits such as improved safety and reduced transportation costs, they also raise concerns regarding job security and reemployment.
- **Customer service:** Chatbots and virtual assistants are becoming increasingly prevalent in customer service roles. These automated systems can handle a wide range of customer inquiries, reducing the need for human customer service representatives. While human interaction may still be necessary for complex or sensitive issues, the deployment of chatbots is likely to reduce the overall demand for customer service jobs.

Addressing the challenges

The displacement effect of technology on certain job sectors raises important challenges that need to be addressed. Here are a few strategies to mitigate the negative impacts and facilitate a smooth transition:

- **Upskilling and retraining:** Investing in education and training programs can empower workers to acquire new skills and adapt to changing job requirements. By aligning training initiatives with emerging technologies, individuals can enhance their employability and transition into new roles that complement technological advancements.

- **Supporting entrepreneurship and innovation:** Encouraging entrepreneurial activities and promoting innovation can create new job opportunities in emerging sectors. By fostering a culture of innovation, governments and organizations can harness the potential of technology to drive economic growth and job creation.

- **Enhancing social safety nets:** Strengthening social safety nets, such as unemployment benefits and healthcare provisions, can provide a safety net for individuals affected by technological displacement. It is crucial to ensure that workers have access to necessary support systems during periods of job transition and uncertainty.

- **Promoting collaboration between humans and technology:** Rather than replacing humans entirely, technology can be leveraged to augment human capabilities. By focusing on human-machine collaboration, organizations can design work processes that effectively combine the strengths of both humans and technology, leading to improved productivity and job satisfaction.

Real-world example

A notable example of the displacement effect can be observed in the manufacturing sector. The automotive industry has witnessed significant automation in recent decades, with robots replacing human workers in various production processes. For instance, in the assembly line of automobile manufacturing plants, robots are adept at performing repetitive tasks with high precision and speed, reducing the need for manual labor. While this automation has led to increased efficiency and cost savings for car manufacturers, it has resulted in job losses for many assembly line workers.

To address these challenges, some companies have implemented programs to reskill and retrain their workforce. By providing opportunities for workers to enhance their technical skills or transition to new roles within the organization, these companies are mitigating the negative effects of technological displacement. Additionally, governments and educational institutions can play a crucial role in providing resources and support for affected workers to acquire new skills and find alternative employment opportunities.

Conclusion

The displacement effect of technology on certain job sectors is an inevitable consequence of technological advancements. While it presents challenges such as job losses and the need for skill adaptation, it also offers opportunities for innovation and increased productivity. By implementing strategies such as upskilling, supporting entrepreneurship, enhancing social safety nets, and promoting human-technology collaboration, we can navigate the changing landscape of work and create inclusive and resilient economies. It is crucial to strike a balance between technological progress and the well-being of workers to ensure a sustainable and equitable future.

Understanding the role of technology in creating new job opportunities

In today's rapidly changing world, technology plays a crucial role in shaping the job market and creating new employment opportunities. Technological advancements have revolutionized industries, disrupted traditional business models, and given rise to entirely new sectors. This section aims to explore the various ways in which technology contributes to the creation of new jobs.

The Impact of Automation and Artificial Intelligence

One of the key drivers of job creation through technology is the automation of routine tasks using artificial intelligence (AI) and advanced technologies. While some fear that automation will lead to widespread job losses, it is important to recognize that it also creates new types of employment.

Automation allows companies to streamline their operations, increase efficiency, and reduce costs. This, in turn, enables businesses to expand their operations, develop new products or services, and enter new markets. As a result, there is a growing demand for skilled workers who can operate, maintain, and develop new technologies. For example, the introduction of automated production lines in manufacturing plants has not only eliminated repetitive and mundane tasks but has also created a need for technicians and engineers with expertise in robotics and automation.

Furthermore, AI technologies are increasingly being used to augment human capabilities rather than replace them entirely. For instance, chatbots and virtual assistants are being employed to handle customer inquiries, freeing up human employees to focus on more complex and strategic tasks. This creates new job opportunities in fields such as customer service, data analysis, and algorithm development.

Emerging Technologies and Job Creation

The rapid advancement of various emerging technologies is giving rise to entirely new job sectors. These technologies include, but are not limited to, the Internet of Things (IoT), virtual reality (VR), augmented reality (AR), blockchain, and 3D printing. Companies investing in these domains are seeking to gain a competitive edge by leveraging the potential of these technologies.

For example, the widespread adoption of IoT is leading to the creation of job roles such as IoT analysts, IoT solutions architects, and IoT security specialists. Similarly, the development of AR and VR technologies has opened doors for software developers, UX designers, and content creators specializing in these immersive experiences.

Blockchain technology, primarily known for its association with cryptocurrencies, has broader applications in sectors such as supply chain management, finance, and healthcare. As companies explore these applications, professionals with expertise in blockchain development, cybersecurity, and smart contract programming are becoming increasingly sought after.

The introduction of 3D printing technology has disrupted traditional manufacturing processes, leading to the demand for specialists in additive manufacturing, 3D printing design, and materials engineering. These professionals help drive innovation and productivity gains in industries ranging from aerospace and automotive to healthcare and fashion.

Digital Transformation and Job Creation

The digital transformation of industries and sectors is another significant driver of job creation. As organizations embrace digital technologies and strategies, there is a need for workers with digital skills and knowledge to support this transition.

Digital marketing, for instance, has emerged as a critical field, encompassing roles such as social media managers, SEO specialists, and digital content creators. As companies and brands seek to expand their online presence and reach, professionals who can harness digital platforms and analytics tools are in high demand.

Moreover, the shift towards remote work and digital collaboration tools has created opportunities for professionals specializing in remote project management, virtual team coordination, and cybersecurity for remote work environments.

Entrepreneurship and Technology-driven Job Creation

Technology has significantly lowered the barriers to entry for entrepreneurs, enabling individuals to create their own businesses and generate employment opportunities. Startups and small businesses that leverage technology can compete with established companies and disrupt traditional industries.

For instance, the rise of e-commerce platforms, supported by advances in logistics and delivery technologies, has enabled entrepreneurs to start their own online stores and reach customers worldwide. This has created jobs in areas such as website development, online marketing, and customer support.

Furthermore, technology-driven platforms such as ride-sharing services and food delivery apps have allowed individuals to become independent contractors and generate income on a flexible basis. These platforms rely on mobile apps, location-based services, and online payment systems, creating opportunities for app developers, data analysts, and customer service representatives.

Addressing Technological Skills Gap

While technology creates new job opportunities, it also highlights the importance of acquiring relevant skills to remain employable in the market. The rapid pace of technological advancements often outpaces the development of skills in the workforce, leading to a significant skills gap.

To address this gap, it is essential for individuals to continuously upskill and reskill themselves to stay relevant in the job market. Employers also have a responsibility to invest in training and development programs to enhance the digital literacy of their workforce.

Governments and educational institutions play a critical role in developing programs that provide accessible and affordable training in emerging technologies. By equipping individuals with the necessary skills, societies can ensure that technological advancements lead to inclusive economic growth and increased employment opportunities.

Conclusion

Technology, when harnessed effectively, has the potential to create new and exciting job opportunities. Automation, emerging technologies, digital transformation, entrepreneurship, and the need for digital skills are driving the creation of jobs in diverse sectors. However, it is crucial to address the skills gap and foster an environment conducive to continuous learning and adaptation. By doing so, we can fully leverage the potential of technology to create a future of inclusive and sustainable employment opportunities.

Discussing the Potential of Technology in Improving Productivity and Wages

In today's rapidly evolving technological landscape, the potential of technology to improve productivity and wages is a topic of great importance. By leveraging the power of advanced digital tools and automation, businesses can optimize their operations, streamline processes, and enhance their overall efficiency. This, in turn, can lead to increased productivity and higher wages for workers. In this section, we will explore the various ways in which technology can positively impact productivity and wages in different sectors.

Understanding the Relationship between Technology and Productivity

Productivity is a key driver of economic growth and is often defined as the amount of output produced per unit of input. Technology has long been recognized as a catalyst for improving productivity by enabling businesses to produce more goods and services with the same or fewer resources. Here, we will discuss some of the main ways in which technology enhances productivity:

- **Automation and Robotics**: Automation of repetitive tasks through the use of robots and other technologies can significantly increase productivity. By replacing manual labor with machines, businesses can achieve higher output rates, reduce errors, and improve overall efficiency.

- **Data Analytics and Artificial Intelligence (AI)**: Technology allows businesses to harness the power of data analytics and AI algorithms to gain insights, make informed decisions, and optimize processes. By analyzing large datasets, businesses can identify patterns, spot inefficiencies, and implement targeted improvements, leading to enhanced productivity.

- **Collaboration Tools and Communication Technologies**: Advanced collaboration tools, such as project management software, video conferencing platforms, and cloud-based document sharing, enable seamless communication and collaboration among team members, regardless of their geographical locations. This improves coordination, reduces delays, and boosts productivity.

- **Internet of Things (IoT):** IoT devices, which connect everyday objects to the internet, enable real-time data collection and analysis. This allows businesses to monitor and optimize processes, anticipate maintenance needs, and make data-driven decisions to enhance productivity.

By embracing these technologies, businesses can leverage the potential of technology to improve productivity, reduce costs, and create new opportunities for workers.

Implications for Wages and Employment

The positive impact of technology on productivity often translates into higher wages for workers. When businesses become more productive, they can afford to pay higher wages, as their overall profitability increases. Additionally, technology also enhances worker's skills and capabilities, making them more valuable in the job market.

However, it is important to note that the relationship between technology, productivity, and wages is complex and multifaceted. Technological advancements can also lead to job displacement in certain sectors where automation replaces routine tasks. This could potentially lead to wage polarization, with high-skilled workers experiencing wage growth, while low-skilled workers face unemployment or stagnant wages.

To address these challenges, policymakers, businesses, and society as a whole need to adopt strategies that ensure technology-driven productivity improvements benefit all workers. This can be achieved through the following measures:

- **Investment in Skills Development:** By investing in training and education programs, workers can build the skills needed to adapt to technological changes. This can help workers stay relevant in the job market and secure higher-paying jobs.

- **Enhancing Labor Market Flexibility:** Policies that promote labor market flexibility, such as supporting entrepreneurship and reducing regulatory barriers, can help create new job opportunities and encourage innovation. This promotes overall wage growth and fosters inclusive economic growth.

- **Rethinking Social Safety Nets:** As technology disrupts traditional job markets, it is crucial to establish social safety nets that provide support to individuals affected by technological advancements. This includes measures like retraining programs, income support, and ensuring access to affordable healthcare and other basic necessities.

By implementing these strategies, we can ensure that the potential of technology to improve productivity also translates into higher wages and greater economic prosperity for all.

Real-World Examples

To further illustrate the potential of technology to improve productivity and wages, let's consider a few real-world examples:

1. **Manufacturing Sector:** The adoption of advanced manufacturing technologies, such as robotics, has significantly increased productivity in the manufacturing sector. For instance, automobile manufacturers have embraced automation in their assembly lines, leading to higher output rates and increased wages for workers who now oversee and maintain the robots.

2. **Retail Sector:** The implementation of data analytics and AI technologies in the retail sector has enhanced inventory management, demand forecasting, and customer targeting. This has improved efficiency, reduced costs, and allowed retailers to offer higher wages to skilled workers who manage these technologies.

3. **Information Technology (IT) Sector:** The IT sector itself is a prime example of how technology can create high-wage jobs. The continual advancements in software development, cybersecurity, and AI have fueled the demand for highly skilled IT professionals. As a result, wages in the IT sector have been consistently high, attracting talented individuals and driving innovation.

These examples demonstrate how technology can positively impact productivity and wages across various industries, leading to overall economic growth.

Challenges and Considerations

While the potential of technology to improve productivity and wages is promising, it is essential to address certain challenges and considerations:

- **Growing Skills Gap:** Rapid technological advancements require a highly skilled workforce. However, there is a growing skills gap in many countries, making it challenging for workers to adapt to the changing demands of the job market. Bridging this gap through targeted education and training programs is crucial to maximize the benefits of technology.

- **Economic Disparities:** Technology has the potential to exacerbate economic disparities if not properly managed. Low-skilled workers may face job displacement, and income inequality can widen. To ensure the benefits of technology are shared equitably, it is important to implement policies that support inclusive economic growth.

- **Ethical Considerations:** The adoption of technologies such as AI and automation raises ethical concerns. It is essential to navigate these considerations and ensure that technology is used responsibly, respects workers' rights, and does not perpetuate biases or discrimination.

Addressing these challenges requires a holistic approach involving collaboration between governments, businesses, and society.

Conclusion

Technology has immense potential to improve productivity and wages by enabling businesses to optimize operations, make informed decisions, and enhance efficiency. However, realizing this potential requires proactive efforts to bridge the skills gap, address economic disparities, and navigate the ethical implications of technological advancements. By embracing a balanced approach, we can harness the power of technology to create inclusive and prosperous economies where the benefits are shared by all.

Investigating the Impact of Automation on Workforce Composition and Skills Requirements

Automation, the use of technology to perform tasks previously done by humans, has had a significant impact on the composition of the workforce and the skills required in various industries. In this section, we will explore how automation has transformed the labor market, examine the skills that are in demand due to automation, and discuss the implications of these changes.

The Transformation of the Labor Market

Automation has led to a transformation in the labor market by changing the types of jobs available and the skills required to perform them. Some jobs that were once done by humans have been completely replaced by machines, while others have been augmented by automation. This shift has resulted in changes in the workforce composition and the skill sets needed to thrive in the new job market.

One clear impact of automation is the displacement effect on certain job sectors. Routine, repetitive tasks that can be easily automated, such as assembly line work or data entry, have been taken over by machines. This has led to a decline in employment opportunities in these sectors, as fewer workers are needed to perform these tasks.

On the other hand, automation has also created new job opportunities. As machines take over routine tasks, humans are freed up to focus on more complex, creative, and non-routine work. For example, in the manufacturing industry, where robots have taken over many repetitive tasks, there is a growing demand for workers with skills in programming, maintenance, and quality control.

Skills in Demand

Automation has reshaped the skills required in the labor market. While routine manual and cognitive skills are becoming less in demand, specific skills related to technology, critical thinking, and interpersonal communication are increasingly sought after.

1. Technological Skills: With automation becoming more prevalent, there is a growing need for workers who can understand and operate the technology. Skills in programming, data analysis, and machine learning are becoming essential for many industries. Proficiency in using digital tools and software is increasingly important across various job roles.

2. Critical Thinking and Problem Solving: As machines take over routine tasks, human workers are expected to focus on more complex problems. Strong critical thinking skills, including the ability to analyze data, make decisions, and solve problems creatively, are highly valued in the automation era.

3. Interpersonal and Communication Skills: While machines can handle many technical tasks, jobs that require human interaction and emotional intelligence are still in demand. Effective communication, collaboration, and empathy are skills that are difficult to automate and are therefore highly valued in various sectors.

4. Adaptability and Lifelong Learning: With automation continuously advancing, workers must be adaptable and willing to learn new skills throughout their careers. The ability to adapt to new technologies and ways of working is essential to thrive in an automated workforce.

Implications and Strategies

The impact of automation on workforce composition and skills requirements has both positive and negative implications. On one hand, automation has the potential to increase productivity, improve safety, and create new job opportunities. On the other hand, it can lead to job displacement and exacerbate inequality if workers lack the necessary skills to transition into new roles.

To address the challenges posed by automation, it is crucial to invest in upskilling and reskilling programs. These initiatives should focus on providing workers with the skills needed to thrive in a technology-driven work environment. This can be done through vocational training, apprenticeships, online learning platforms, and collaborations between educational institutions and industries.

Furthermore, governments and organizations should promote a culture of lifelong learning, encouraging individuals to continuously update their skills to remain relevant in the labor market. Additionally, policies that support social safety nets, such as unemployment benefits and income support, can help workers transition during periods of job displacement.

In summary, the impact of automation on workforce composition and skills requirements has fundamentally changed the labor market. As routine tasks are automated, the demand for technological, critical thinking, and interpersonal skills has increased. To navigate the automation era successfully, individuals and society as a whole must adapt and invest in skills development and lifelong learning. This will ensure the workforce remains resilient and prepared for the challenges and opportunities presented by automation.

Assessing the role of government policies in addressing the challenges of technology and employment

Government policies play a crucial role in addressing the challenges that arise from the intersection of technology and employment. As technology continues to advance at a rapid pace, it brings both opportunities and challenges in the labor market. While technology has the potential to create new jobs and increase productivity, it also poses the risk of job displacement and widening income inequalities. In this section, we will assess the role of government policies in addressing these challenges and creating a favorable environment for technology and employment to coexist harmoniously.

Understanding the challenges of technology and employment

Before discussing the role of government policies, it is important to understand the challenges that technology poses to employment. One of the main challenges is the displacement effect, wherein technological advancements replace certain job sectors, leading to job losses. Automation, artificial intelligence, and machine learning are increasingly replacing routine and repetitive tasks, which primarily affect low-skilled workers.

Another challenge is the changing nature of jobs due to technology. As technology reshapes industries, there is a shift in the skills required by the workforce. This can create skill gaps, leading to unemployment or underemployment if workers do not possess the necessary skills for the emerging jobs.

Additionally, technology-driven globalization enables companies to outsource jobs to countries with lower labor costs, which can further exacerbate unemployment in developed economies. This offshoring of jobs can result in economic and social inequalities within and between countries.

Government policies for addressing technology and employment challenges

To address the challenges of technology and employment, governments can implement various policies to ensure that the benefits of technological advancements are shared by all and that the workforce is adequately prepared for the changing job landscape. The following are some key policy interventions that governments can consider:

1. **Education and training policies:** As technology continues to advance, the need for an adaptive and skilled workforce becomes crucial. Governments can invest in education and training programs that focus on equipping individuals with the necessary skills for the jobs of the future. This includes promoting science, technology, engineering, and mathematics (STEM) education, providing vocational training programs, and facilitating lifelong learning opportunities to enable individuals to upskill or reskill as needed.

2. **Labor market and employment policies:** Governments can implement labor market policies that support a smooth transition for workers affected by technological disruptions. This includes providing unemployment benefits, job counseling, and reemployment services to support workers who lose their jobs due to technological advancements. Additionally, governments can facilitate job matching platforms that connect displaced workers with new employment opportunities.

3. **Income support policies:** To mitigate the potential income inequalities arising from technology-driven job displacements, governments can implement income

support policies such as minimum wage laws, social safety nets, and income redistribution programs. These policies aim to ensure that the benefits of economic growth generated by technology are shared among all segments of society.

4. **Innovation and entrepreneurship policies:** Governments can promote innovation and entrepreneurship as a means to create new job opportunities. This includes supporting research and development activities, providing financial incentives for startups and small and medium-sized enterprises (SMEs), and fostering an entrepreneurial ecosystem that encourages the creation of innovative businesses.

5. **Collaboration and partnerships:** Governments can collaborate with industry associations, labor unions, and educational institutions to develop policies and strategies that address the challenges of technology and employment. Such partnerships can facilitate knowledge sharing, skill development initiatives, and the identification of emerging job trends, ensuring that policies are effectively aligned with the needs of the labor market.

While these policies can be effective in addressing the challenges of technology and employment, it is important for governments to strike a balance between facilitating technological advancements and ensuring social inclusiveness. They should also consider the ethical implications of policy interventions, particularly regarding privacy, data protection, and algorithmic biases.

Case study: Government policies in Sweden

Sweden serves as a notable case study in addressing the challenges of technology and employment through government policies. The Swedish government has implemented a comprehensive approach that focuses on lifelong learning, active labor market policies, and social welfare programs.

One key policy initiative is the Swedish Public Employment Service, which provides job counseling, training, and support services to individuals affected by technological disruptions. This helps displaced workers to reskill, upskill, or find new employment opportunities. The government also emphasizes collaboration between employers, trade unions, and educational institutions to ensure that education and training programs align with the changing demands of the labor market.

Furthermore, Sweden has adopted policies that encourage innovation and entrepreneurship. The government provides financial support for startups and supports incubators and accelerators to foster a favorable environment for technological innovation. These policies have resulted in the growth of innovative companies and the creation of new job opportunities.

The Swedish government also places a strong emphasis on income redistribution and social welfare programs to address income inequalities. The country has implemented a progressive income tax system and generous social security benefits, ensuring a high level of social protection for its citizens.

Conclusion

Government policies play a vital role in addressing the challenges of technology and employment. By implementing comprehensive policies that focus on education and training, labor market interventions, income support, innovation and entrepreneurship, and collaborative partnerships, governments can create an environment that facilitates the integration of technology and employment. The case study of Sweden highlights the effectiveness of such policies in addressing the challenges and creating a favorable environment for the workforce to adapt and thrive in the era of technological advancements.

As technology continues to evolve, governments must continuously monitor and adapt their policies to keep pace with the changing dynamics of the labor market. By fostering inclusive and responsible technology-driven economies, governments can ensure that the benefits of technology are harnessed while minimizing negative impacts on employment and inequalities.

Examining the concept of digital divide and its implications on employment opportunities

The concept of the digital divide refers to the gap between individuals and communities that have access to digital technologies and those who do not. This divide encompasses disparities in internet access, computer literacy, and availability of technology infrastructure. In today's interconnected world, the digital divide holds significant implications for employment opportunities.

Understanding the digital divide

The digital divide can manifest in various forms, including:

- **Access divide:** This refers to the unequal distribution of internet access and technological resources. In many developing countries and marginalized communities, limited infrastructure and high costs hinder access to broadband internet connections.

- **Usage divide:** Even when access is available, differences in individuals' ability to effectively use technology can create a usage divide. This can be due to low digital literacy, lack of training, or language barriers.

- **Empowerment divide:** The empowerment divide relates to the unequal access to opportunities and benefits that digital technologies bring. This can include limited access to online education, job portals, and e-commerce platforms.

Implications on employment opportunities

The digital divide has profound implications for employment opportunities, both at individual and societal levels. Here are some key factors to consider:

1. **Limited access to job markets:** Lack of internet access and digital skills can hinder individuals' ability to search and apply for job opportunities. Many employers now rely on online platforms for recruitment, making it challenging for those without internet access to enter the job market.

2. **Skills mismatch:** As technology becomes increasingly integrated into the workplace, digital skills are becoming essential for many jobs. The digital divide exacerbates the skills mismatch, as individuals with limited access to technology may not acquire the necessary skills and fall behind in the job market.

3. **Increasing automation:** Automation and artificial intelligence are transforming industries, with the potential to replace certain jobs. Those who lack digital skills and access to technology are particularly vulnerable to job displacement in sectors undergoing technological advancements.

4. **Reproduction of inequalities:** The digital divide can perpetuate existing social and economic inequalities. Marginalized communities and individuals with limited resources are more likely to face barriers in accessing technology, resulting in limited employment opportunities and a widening income gap.

Addressing the digital divide

Addressing the digital divide is crucial for fostering inclusive and equitable employment opportunities. Here are some strategies to consider:

1. **Infrastructure development:** Governments and organizations should invest in expanding technological infrastructure, especially in underserved areas. This includes improving internet connectivity and ensuring affordable access to technology.

2. **Digital literacy programs:** Initiatives should be launched to provide training and education on digital skills. These programs can empower individuals

with the necessary knowledge to navigate the online job market and adapt to evolving technology requirements.

3. **Public-private partnerships:** Collaboration between governments, private sector entities, and non-profit organizations can facilitate resource sharing and investment in digital inclusion initiatives. Such partnerships can help bridge the digital divide and foster employment opportunities.

4. **Promoting digital entrepreneurship:** Encouraging the growth of digital entrepreneurship can create job opportunities for individuals with limited access to traditional employment. Supporting start-ups and providing entrepreneurial training can empower individuals and communities to leverage technology for economic growth.

Case study: The One Laptop per Child (OLPC) initiative

The One Laptop per Child initiative is an example of an effort to bridge the digital divide and enhance employment opportunities in disadvantaged communities. The initiative aimed to provide low-cost laptops to children in developing countries, promoting digital literacy and educational opportunities. By equipping young individuals with digital skills from an early age, the initiative aimed to prepare them for future employment opportunities.

The OLPC initiative demonstrated the potential of technology in narrowing the digital divide and fostering employment prospects. However, it also highlighted the importance of addressing infrastructure challenges and ensuring the sustainability of such programs.

Conclusion

The digital divide poses significant challenges to employment opportunities in today's technology-driven world. Addressing this divide is crucial for promoting inclusive economic growth and reducing inequalities. Through investments in infrastructure, digital literacy programs, and collaborative efforts, societies can create a more equitable and sustainable future where technology is accessible to all. Only by bridging the digital divide can individuals and communities fully harness the potential of technology for employment and economic empowerment.

Understanding the relationship between technology and income inequality

In this section, we will examine the relationship between technology and income inequality. We will explore how technological advancements can both exacerbate and alleviate income inequality. We will also discuss the factors that contribute to this relationship and the potential strategies for reducing income inequality in the digital age.

The Impact of Technological Advancements on Income Inequality

Technological advancements have the potential to impact income inequality in various ways. On the one hand, technology can increase income inequality by creating a skills gap. As certain jobs become automated, workers with obsolete skills may experience job displacement and reduced wages. This can lead to an increase in income inequality between those who have the necessary skills to thrive in the digital economy and those who do not.

On the other hand, technology can also reduce income inequality by creating opportunities for innovation and entrepreneurship. The digital age has democratized access to information and tools, enabling individuals from diverse backgrounds to create new businesses and generate income. Technological advancements have also led to the creation of new industries and jobs, providing avenues for economic advancement.

Factors Contributing to the Relationship between Technology and Income Inequality

Several factors contribute to the relationship between technology and income inequality.

One important factor is the digital divide. The digital divide represents the gap in access to information and communication technologies, such as the internet, between different socioeconomic groups. Limited access to technology can hinder individuals' ability to acquire necessary skills and participate fully in the digital economy, exacerbating income inequality.

Another factor is the skill-biased nature of technological change. Certain technologies may require specific skills, such as programming or data analysis, which are in high demand in the digital economy. If individuals do not have access to quality education and training opportunities to acquire these skills, they may be left behind, further widening income disparities.

Additionally, the concentration of economic power in technology firms can contribute to income inequality. As technology companies become increasingly dominant in their respective industries, they can exert significant influence over wages and working conditions. This concentration of power can lead to income disparities between workers in the technology sector and those in other sectors of the economy.

Strategies for Reducing Income Inequality in the Digital Age

Addressing income inequality in the digital age requires a multi-faceted approach. Here are some strategies that can help reduce income disparities:

1. Enhancing digital literacy and access: Promoting digital literacy programs and initiatives can help bridge the digital divide. Providing affordable access to technology and the internet can ensure equal opportunities for individuals to acquire necessary digital skills.

2. Investing in education and lifelong learning: Fostering a culture of lifelong learning and ensuring access to quality education can equip individuals with the skills needed to thrive in the digital economy. Investing in science, technology, engineering, and mathematics (STEM) education can help prepare future generations for technology-driven careers.

3. Promoting inclusive innovation ecosystems: Creating innovation ecosystems that are inclusive and diverse can help address income inequality. Supporting entrepreneurship among underrepresented groups, such as women and minorities, can foster greater equality in the benefits of technological advancements.

4. Implementing progressive taxation and redistribution policies: Progressive taxation can help redistribute wealth and reduce income disparities. Revenues generated from progressive taxation can be used to fund social programs that provide support and opportunities for disadvantaged individuals and communities.

5. Strengthening worker protections and rights: Ensuring fair wages, safe working conditions, and protection of workers' rights can help reduce income inequality. Policies that promote collective bargaining and protect workers from exploitation can help create a more equitable labor market.

6. Encouraging responsible corporate practices: Encouraging technology companies to adopt responsible business practices, such as fair labor standards and equitable distribution of profits, can contribute to reducing income inequality.

7. Fostering international collaboration: Promoting international collaboration in technology and innovation can help reduce global income inequalities. Sharing knowledge, resources, and technology with developing countries can enable them to leapfrog to more advanced stages of economic development.

In conclusion, the relationship between technology and income inequality is complex and multifaceted. While technology can exacerbate income disparities, it also has the potential to reduce inequality through innovation and entrepreneurship. By addressing the factors that contribute to income inequality and implementing targeted strategies, we can strive towards a more equitable and inclusive digital economy.

Analyzing the role of education and skills development in adapting to technological changes

As technological advancements continue to reshape industries and economies, the role of education and skills development becomes increasingly crucial. In this section, we will explore the importance of education and skills in adapting to technological changes, and how individuals and societies can prepare themselves to thrive in a rapidly evolving technological landscape.

The Need for Continuous Learning

In today's fast-paced world, the half-life of skills is rapidly shrinking. This means that the knowledge and skills acquired through formal education might become obsolete in a short span of time. Therefore, individuals must embrace the idea of lifelong learning and actively engage in continuous upskilling and reskilling.

To adapt to technological changes, individuals need to develop a growth mindset that embraces learning and embraces change. This mindset allows them to be open to new ideas, eager to explore emerging technologies, and willing to acquire new skills.

Building a Foundation of Digital Literacy

As technology becomes increasingly pervasive, digital literacy has become an essential skill set for individuals in almost every field. Digital literacy refers to the ability to use, understand, and critically evaluate digital technologies effectively.

Educational institutions play a vital role in equipping individuals with digital literacy skills. They need to provide comprehensive training in areas such as computer literacy, internet usage, information management, and digital communication. By developing strong digital literacy skills, individuals can navigate the digital landscape, adapt to new technologies, and leverage them for personal and professional growth.

Adapting Curricula to Technological Changes

Education systems must adapt to technological changes by updating their curricula to reflect the evolving needs of industries and the job market. This involves a close collaboration between academic institutions, industry experts, and policymakers.

A curriculum that integrates technology-related subjects and practical hands-on experiences can help bridge the gap between education and the real-world applications of technology. Additionally, incorporating interdisciplinary approaches that foster problem-solving, critical thinking, and creativity can better prepare individuals to navigate complex technological environments.

Promoting STEM Education

Science, technology, engineering, and mathematics (STEM) education plays a crucial role in preparing individuals for careers in technology-driven industries. By nurturing an interest in these fields from a young age, educational institutions can develop a strong foundation of STEM skills that can be built upon later in life.

To encourage STEM education, schools should promote hands-on learning experiences and provide access to relevant technology and equipment. Additionally, initiatives that aim to increase diversity and inclusion in STEM fields can help ensure that the benefits of technological advancements are accessible to all.

Fostering Collaboration and Problem-Solving Skills

Technological changes often require individuals to work collaboratively to solve complex problems. Therefore, education systems should emphasize the development of collaboration and problem-solving skills.

By fostering a collaborative learning environment, educational institutions can encourage teamwork, communication, and interdisciplinary thinking. Group projects, case studies, and interactive activities can provide opportunities for students to tackle real-world problems and develop these essential skills.

Addressing the Skills Gap

As technology continues to advance, there is a growing concern about the skills gap - the mismatch between the skills demanded by the job market and those possessed by the workforce. To address this gap, educational institutions need to provide targeted training and education programs that align with the evolving needs of industries.

Partnerships between educational institutions and industry stakeholders can help bridge the skills gap by designing curriculum and training programs that are directly informed by industry demands. By aligning education with industry needs, individuals can acquire the skills necessary to adapt to technological changes and secure meaningful employment.

Life-Long Learning as a Career Investment

In the face of rapid technological changes, individuals need to view lifelong learning as a long-term career investment. By actively seeking out learning opportunities, individuals can remain competitive and expand their career prospects.

Continuous learning can take various forms, including online courses, workshops, conferences, certifications, and professional development programs. It is essential for individuals to create a personal learning roadmap and take proactive steps to acquire new skills and knowledge throughout their careers.

Real-World Application: Upskilling for Industry 4.0

Industry 4.0, characterized by the integration of digital technologies and automation in manufacturing, presents both opportunities and challenges for the workforce. To adapt to this new era, individuals must upskill themselves and acquire the necessary knowledge and skills.

For example, in the context of Industry 4.0, individuals can benefit from developing proficiency in areas such as data analytics, artificial intelligence, robotics, and cybersecurity. By gaining expertise in these emerging fields, individuals can position themselves as valuable assets in the job market.

Conclusion

Education and skills development play a critical role in adapting to technological changes. Lifelong learning, digital literacy, collaboration, problem-solving skills, and addressing the skills gap are all essential components of preparing individuals and societies for the challenges and opportunities brought about by technological advancements.

By embracing continuous learning and equipping individuals with the necessary skills, education systems can empower individuals to thrive in a technology-driven world and contribute to the economic growth and development of their communities.

Discussing the potential of technology in reducing gender and minority inequalities

In recent years, there has been a growing recognition of the need to address gender and minority inequalities in various aspects of society, including the realm of technology. Technology, if leveraged effectively, has the potential to significantly reduce these inequalities and foster inclusivity and diversity. In this section, we will explore the ways in which technology can be a catalyst for positive change in terms of gender and minority empowerment.

Understanding the gender and minority inequalities

To fully comprehend the potential of technology in reducing gender and minority inequalities, it is crucial to first gain an understanding of the nature and extent of these inequalities. Gender inequality refers to the unequal treatment and opportunities experienced by individuals based on their gender, often resulting in disparities in education, employment, and income. Minority inequalities, on the other hand, encompass a wide range of disadvantages faced by individuals from minority groups, which may include racial, ethnic, religious, or cultural differences.

These inequalities persist in various domains, including access to education, job opportunities, wage gaps, and underrepresentation in leadership positions. Addressing these issues requires a multi-faceted approach that involves societal changes, policy interventions, and the use of technology as an enabler.

The role of technology in reducing gender and minority inequalities

1. Bridging the digital divide: One of the most crucial steps in reducing gender and minority inequalities is ensuring equal access to technology and the internet. Technology can play a key role in bridging the digital divide by providing affordable and accessible devices, internet connectivity, and digital literacy training to marginalized communities. Initiatives such as community technology centers and public Wi-Fi networks can help empower individuals with the necessary skills and resources to participate fully in the digital age.

2. Enhancing educational opportunities: Technology can revolutionize education and provide equal opportunities for all learners. Online learning platforms, educational apps, and digital resources can break down barriers to education faced by marginalized groups, who may have limited access to quality schools or face social, economic, or geographical constraints. By providing flexible and personalized learning experiences, technology can empower individuals to acquire knowledge and skills, regardless of their gender or minority status.

3. Promoting inclusive workplaces: Technology can foster inclusive workplaces by mitigating biases, facilitating remote work, and providing equal opportunities for career advancement. For example, algorithms can remove gender-biased language from job descriptions, ensuring equal representation in the recruitment process. Remote work facilitated by technology can help individuals with caregiving responsibilities or physical disabilities to pursue meaningful

employment. Additionally, technology-driven mentorship programs and networking platforms can connect minority professionals with industry leaders, fostering diversity in leadership positions.

4. Empowering entrepreneurship: Technology has the potential to level the playing field for aspiring entrepreneurs from diverse backgrounds. Online platforms and e-commerce websites enable individuals to start their businesses with minimal upfront costs and reach a global audience. Technology can also facilitate access to financial services, such as crowdfunding platforms and digital payment systems, which are essential for starting and growing businesses. By empowering entrepreneurship, technology can create economic opportunities and reduce inequalities faced by gender and minority groups.

5. Addressing bias and discrimination: While technology itself is neutral, it can reflect and perpetuate existing biases and discrimination. However, when used intentionally, technology can help address these issues. For example, algorithms can be designed to minimize bias in hiring processes and eliminate discriminatory practices. Open-source technologies and decentralized platforms can provide opportunities for underrepresented groups to develop and showcase their skills and talents.

Case study: TechGirls initiative

The TechGirls initiative is an exemplary case that showcases the potential of technology in reducing gender inequalities. The initiative, launched by the U.S. Department of State, aims to empower young women from the Middle East and North Africa (MENA) region through an intensive exchange program focused on science, technology, engineering, and math (STEM).

Through the TechGirls program, girls from underprivileged backgrounds are given the opportunity to engage in hands-on workshops, site visits, mentorship, and collaborative projects with American counterparts. The program not only exposes the participants to technology and innovation but also fosters cross-cultural understanding and collaboration.

By leveraging technology as a means of empowerment, the TechGirls initiative has successfully addressed gender inequalities faced by young women in the MENA region. The program has inspired and equipped a new generation of female leaders in STEM fields and has had a ripple effect in promoting gender equality on a larger scale.

Challenges and considerations

While the potential of technology in reducing gender and minority inequalities is promising, several challenges need to be addressed:

1. Access and affordability: The digital divide continues to persist, with marginalized communities still lacking access to affordable devices and reliable internet connectivity. Efforts must be made to provide universal access to technology, ensuring that cost and infrastructure barriers do not further exacerbate inequalities.

2. Bias and discrimination: The design and implementation of technology solutions must be mindful of the risks of perpetuating biases and discrimination. Ethical considerations and diversity in technology teams are essential to mitigate these risks and ensure inclusive outcomes.

3. Digital skills and literacy: Merely providing access to technology is not enough. Efforts must be made to equip individuals with the necessary digital skills and literacy required to fully participate in the digital era. Technology training programs and digital literacy initiatives should be tailored to the needs of diverse communities.

4. Intersectionality: Gender and minority inequalities intersect with other forms of discrimination, such as socioeconomic status and disability. Solutions aimed at reducing inequalities must take into account these intersecting factors to ensure holistic and inclusive outcomes.

Conclusion

Technology has the potential to be a powerful force for reducing gender and minority inequalities. By bridging the digital divide, enhancing educational opportunities, promoting inclusivity in workplaces, empowering entrepreneurship, and addressing biases, technology can create a more equitable and diverse society. However, this potential can only be realized through concerted efforts, policy interventions, and ethical considerations. It is imperative that technology be leveraged as a tool for positive change, fostering an inclusive and equitable future for all.

Exploring strategies for fostering inclusive and equitable technology-driven economies

In today's increasingly technology-driven world, it is crucial to ensure that the benefits of technological advancements are shared equitably and inclusively. This section will discuss strategies for fostering inclusive and equitable technology-driven economies. We will explore various approaches that can help bridge the digital divide, promote equal access to technology and opportunities, and address the challenges of inequality in a rapidly evolving technological landscape.

Understanding the Digital Divide

Before delving into strategies, it is essential to understand the concept of the digital divide. The digital divide refers to the gap between those who have access to digital technologies and those who do not. This divide is often influenced by factors such as income, education, geographic location, gender, and age. Bridging the digital divide is crucial for fostering inclusive and equitable technology-driven economies.

Education and Digital Skills Development

One of the key strategies for fostering inclusive technology-driven economies is to focus on education and digital skills development. Providing access to quality education that emphasizes digital literacy and technology skills is essential. This can be achieved through targeted initiatives that provide training, resources, and support to individuals, especially those from disadvantaged backgrounds. By equipping individuals with the necessary digital skills, we can enhance their employability and empower them to participate effectively in the digital economy.

Promoting Access to Technology and Infrastructure

Another critical aspect of fostering inclusive technology-driven economies is ensuring equal access to technology and infrastructure. This can be achieved by implementing policies and initiatives that aim to bridge the digital divide. For instance, governments can invest in expanding broadband connectivity to underserved areas, especially in rural and remote regions. Additionally, providing subsidies or affordable options for internet access and technology devices can enable more individuals to access and benefit from technology.

Supporting Entrepreneurship and Innovation

Promoting entrepreneurship and innovation among underrepresented groups is crucial for fostering inclusive and equitable technology-driven economies. Governments and organizations can create initiatives that provide financial support, mentorship, and networking opportunities to aspiring entrepreneurs from marginalized communities. By fostering diversity in entrepreneurship, we can tap into a wider range of perspectives, ideas, and talents, leading to more inclusive and innovative solutions.

Addressing Bias and Discrimination

To foster inclusive and equitable technology-driven economies, it is essential to address bias and discrimination that may exist within the technology industry. This requires promoting diversity and inclusion in tech companies and organizations. Implementing policies that encourage equal opportunity employment, diverse hiring practices, and inclusive workplace environments can help create a more equitable technology sector. Additionally, addressing algorithmic bias in AI systems and ensuring fairness in technology applications can contribute to reducing inequality and promoting inclusivity.

Promoting Collaboration and Partnerships

Collaboration and partnerships among governments, private sector companies, academia, and civil society organizations are instrumental in fostering inclusive and equitable technology-driven economies. By working together, these stakeholders can pool resources, expertise, and knowledge to develop and implement innovative solutions. Public-private partnerships, research collaborations, and cross-sectoral initiatives can drive inclusive growth and ensure that technological advancements are harnessed for the benefit of all.

Case Studies and Best Practices

Analyzing successful case studies and best practices can provide valuable insights into effective strategies for fostering inclusive technology-driven economies. By studying examples from different regions and sectors, we can identify approaches that have yielded positive outcomes in terms of reducing inequality, promoting equal access, and creating opportunities for all. These case studies can serve as inspiration and provide guiding principles for policymakers, educators, entrepreneurs, and other stakeholders involved in driving inclusive and equitable technology-driven economies.

Conclusion

Fostering inclusive and equitable technology-driven economies is a multifaceted endeavor that requires the concerted efforts of various stakeholders. By focusing on education, promoting access to technology and infrastructure, supporting entrepreneurship and innovation, addressing bias and discrimination, fostering collaboration and partnerships, and learning from successful case studies and best practices, we can work towards a future where technological advancements benefit everyone and contribute to inclusive and sustainable economic growth. It is an ongoing process that requires continuous evaluation, adaptation, and collaboration to ensure that technology is harnessed as a force for positive change.

Technology and Globalization

Analyzing the impact of technology on globalization

Understanding the role of technology in driving international trade and investment

In today's interconnected world, technology plays a crucial role in facilitating international trade and attracting foreign investment. The advancements in digital technologies, telecommunications, and transportation infrastructure have significantly transformed the global marketplace, breaking down geographical barriers and enabling businesses to expand their reach beyond national borders. This section will explore the various ways in which technology drives international trade and investment, and how it has become an essential driver of economic growth and development.

The role of technology in expanding market access

One of the fundamental ways in which technology stimulates international trade is by expanding market access for businesses. With the advent of the internet and e-commerce platforms, companies can now sell their products and services to customers around the globe without the need for physical presence in foreign markets. Online marketplaces provide a platform for small and medium-sized enterprises (SMEs) to access a global customer base and compete on an equal footing with larger multinational corporations. This has led to increased participation of SMEs in international trade, resulting in greater diversification and efficiency in global supply chains.

Technology has also enabled businesses to engage in cross-border transactions more seamlessly. Electronic payment systems, such as mobile payment apps and cryptocurrencies, have made it easier and more secure for buyers and sellers to conduct international financial transactions, reducing transaction costs and currency exchange risks. Additionally, digital supply chain management systems allow businesses to track and manage their global production and distribution networks in real-time, facilitating the efficient movement of goods and services across borders.

Enabling innovation and productivity gains

Technological advancements drive international trade by fostering innovation and increasing productivity. Innovation is the engine of economic growth, and technology acts as a catalyst for innovation by enabling businesses to develop new products, services, and production processes. The use of advanced technologies, such as artificial intelligence, machine learning, and big data analytics, helps businesses gain insights into market trends, consumer preferences, and competitive dynamics, enabling them to tailor their offerings to international markets.

Moreover, technology enhances productivity by automating tasks, streamlining processes, and improving operational efficiency. Digitalization and automation of production processes enable businesses to achieve economies of scale, reduce costs, and meet international quality standards. This improved productivity and cost competitiveness enhance a country's ability to export goods and services, attracting foreign buyers and stimulating international trade.

Facilitating global collaboration and networking

Technology has revolutionized communication and collaboration, enabling businesses to connect and network with international partners, suppliers, and customers. Through video conferencing, email, and instant messaging platforms, businesses can now communicate and negotiate with partners across different time zones, overcoming geographical and cultural barriers. This ease of communication encourages collaboration, knowledge sharing, and innovation across borders, fostering the creation of global value chains.

Global networking platforms, such as social media and online professional networks, provide businesses with opportunities to establish and maintain relationships with international customers and suppliers. These platforms enable businesses to showcase their products and services, conduct market research, and explore potential business partnerships. By leveraging technology, businesses can build trust and establish lasting relationships, critical for successful international trade and investment.

Addressing trade and investment barriers

Technology also plays a crucial role in addressing traditional barriers to trade and investment. Digitization of trade documents, such as electronic invoicing and digital certificates of origin, simplifies administrative procedures, reduces trade costs, and facilitates customs clearance processes. This streamlines international transactions and reduces the time and costs associated with international trade.

Furthermore, technology fosters transparency and reduces information asymmetry in international trade. Online platforms that provide access to product information, certifications, and customer reviews empower buyers to make informed decisions and trust the quality and origin of the products they purchase. This transparency nurtures trust between buyers and sellers, facilitating cross-border transactions.

In addition, technology-driven innovations, such as blockchain technology, have the potential to revolutionize international trade and investment by providing secure and transparent transactions, eliminating intermediaries, and reducing transaction costs. Blockchain technology allows for the creation of decentralized and tamper-proof digital ledgers, which can be used to track and verify ownership, authenticity, and compliance of goods and services across borders.

Challenges and opportunities

While technology offers numerous opportunities for driving international trade and investment, it also presents challenges that need to be addressed. One significant challenge is the digital divide, which refers to the gap in internet access and digital skills between countries and within societies. Bridging this divide requires investments in infrastructure, digital literacy, and affordable access to digital technologies to ensure that all countries and individuals can participate in the global digital economy.

Another challenge is data privacy and cybersecurity. As international trade becomes increasingly digital, the protection of personal and business data becomes paramount. Governments and businesses need to implement robust data protection measures and cybersecurity protocols to safeguard sensitive information and ensure trust in digital transactions.

Moreover, the rapid pace of technological advancements poses challenges in keeping up with changing technologies and adapting regulatory frameworks. Governments need to create a conducive environment for technological innovation and entrepreneurship, ensuring that regulatory frameworks are flexible enough to embrace emerging technologies while protecting consumer rights and ensuring fair competition.

In conclusion, technology plays a vital role in driving international trade and investment by expanding market access, fostering innovation and productivity gains, facilitating global collaboration and networking, and addressing traditional trade and investment barriers. Harnessing the potential of technology to overcome challenges and embrace opportunities is crucial for countries and businesses seeking to thrive in the global marketplace.

Examining the globalization of technology industries and supply chains

In today's interconnected world, globalization has become a driving force behind the growth and development of technology industries and supply chains. The rapid advancement of technology and communication networks has made it possible for companies to expand their operations beyond national borders and tap into global markets. This section will examine the various aspects of the globalization of technology industries and supply chains, including its impact, challenges, and opportunities.

Globalization and Technology Industries

Globalization has had a profound impact on technology industries, enabling them to create a global presence and reach a wider customer base. Companies in the technology sector have leveraged advancements in communication technology to establish global manufacturing, research and development (R&D), sales, and service operations.

One key aspect of the globalization of technology industries is the establishment of global technology clusters. These clusters consist of geographically concentrated technology companies, suppliers, and research institutions, fostering collaboration and innovation. For example, Silicon Valley in the United States and Silicon Alley in New York City have emerged as globally recognized technology clusters, housing numerous technology companies and startups.

The globalization of technology industries has also led to the outsourcing of manufacturing operations, especially in the electronics industry. Companies often outsource manufacturing to countries with lower labor costs, such as China and India. This has resulted in the integration of global supply chains, where components and products are sourced from different countries and assembled in a central location.

Globalization and Supply Chains

The globalization of technology industries has had a significant impact on supply chains, transforming them into global networks spanning multiple countries. This globalization of supply chains has been attributed to various factors, including cost reduction, access to markets, and availability of resources.

One of the key drivers of the globalization of supply chains is the pursuit of cost reduction. By sourcing components, raw materials, and finished products from different countries, companies can take advantage of lower production costs and economies of scale. This has resulted in the emergence of complex supply chain networks that span across continents and involve multiple stakeholders.

Another important aspect of the globalization of supply chains is the access to global markets. By establishing a global presence, companies can tap into new markets and diversify their customer base. This not only helps in increasing sales but also mitigates risks associated with dependence on a single market.

Furthermore, globalization has enabled companies to access resources that may not be available domestically. For example, countries rich in natural resources, such as rare earth minerals, can supply these resources to technology companies worldwide. This highlights the interdependence of countries in global supply chains and the need for strategic resource management.

Challenges and Opportunities

While the globalization of technology industries and supply chains presents numerous opportunities, it also brings along several challenges that need to be addressed. One of the key challenges is the management of global supply chains, which are often complex and involve coordination among various stakeholders across different countries. This requires effective communication, logistics management, and risk mitigation strategies.

Moreover, globalization has raised concerns about the exploitation of labor in developing countries. Companies need to ensure that their supply chains adhere to ethical standards and promote fair labor practices. This includes providing safe working conditions, fair wages, and respecting human rights.

Another challenge is the protection of intellectual property rights (IPR) in a globalized environment. Technology industries are often at the forefront of innovation, and the theft or infringement of intellectual property poses a significant risk. Countries need to establish robust IPR laws and enforcement mechanisms to protect the interests of technology companies.

Despite these challenges, the globalization of technology industries and supply chains also presents several opportunities. It allows for the exchange of knowledge and expertise among countries, fostering innovation through collaboration. The globalization of technology industries enables companies to create synergies and leverage complementary capabilities across borders, leading to the development of new products and services.

Additionally, the globalization of supply chains has facilitated the growth of small and medium-sized enterprises (SMEs) by providing them with access to global markets and resources. SMEs can leverage global supply chains to expand their customer base and compete with larger companies on a global scale.

Case Study: The Globalization of the Electronics Industry

The electronics industry provides a compelling case study of the globalization of technology industries and supply chains. Over the past few decades, the electronics industry has witnessed a significant shift in manufacturing and supply chain operations.

Initially, most electronics manufacturing was concentrated in the United States, Japan, and a few European countries. However, with advancements in communication technology and the lure of lower labor costs, manufacturing gradually shifted to East Asian countries, particularly China. Today, China has become a global manufacturing hub for electronics, with numerous multinational companies establishing manufacturing facilities in the country.

The globalization of the electronics industry has resulted in the integration of complex global supply chains. Components and raw materials are sourced from various countries, such as Japan, South Korea, Taiwan, and Southeast Asian nations. These components are then shipped to China for assembly and subsequent distribution to global markets.

The globalization of the electronics industry has not only led to cost reduction but also enabled companies to respond quickly to changes in customer demand. By leveraging global supply chains, companies can adjust production volumes and adapt to market trends, ensuring timely delivery of products to customers worldwide.

However, the electronics industry also faces challenges associated with the globalization of supply chains. This includes issues related to supply chain disruptions, such as natural disasters and geopolitical tensions. Companies need to develop strategies to mitigate these risks and ensure business continuity.

In conclusion, the globalization of technology industries and supply chains has transformed the way companies operate and compete in today's globalized economy. It has created opportunities for innovation, market expansion, and resource access. However, it also poses challenges related to supply chain management, intellectual property protection, and ethical considerations. Understanding the dynamics of globalization in technology industries and supply chains is crucial for companies and policymakers to navigate the complexities of the global economy and harness its potential for sustainable development.

Discussing the challenges and opportunities of technology in developing economies

In developing economies, technology plays a crucial role in driving economic growth and promoting sustainable development. However, these economies face unique challenges and opportunities in harnessing technology for their advancement. In this section, we will explore some of the key challenges and opportunities faced by developing economies in adopting and leveraging technology for their development.

Challenges

1. **Limited Access to Technology:** One of the primary challenges in developing economies is the limited access to technology. Many individuals and communities in these economies lack the necessary infrastructure and resources to access and utilize technology effectively. This digital divide can further exacerbate existing inequalities and hinder economic progress.

2. **Infrastructural Constraints:** Developing economies often face infrastructural constraints, such as inadequate power supply, limited internet connectivity, and underdeveloped transportation networks. These constraints can impede the adoption and diffusion of technology, hampering economic growth and development.

3. **Lack of Skills and Technological Literacy:** Another challenge is the limited availability of skilled individuals with the necessary technical expertise to develop, implement, and maintain technological solutions. Additionally, there is often a lack of technological literacy among the population, making it difficult to fully utilize and benefit from technology.

4. **Financial Limitations:** Developing economies often face financial limitations, making it challenging to invest in and adopt new technologies. Limited access to capital and high costs associated with technology acquisition and infrastructure development can hinder the integration of technology into various sectors.

5. **Institutional and Regulatory Roadblocks:** Inadequate legal and regulatory frameworks, corruption, and bureaucratic inefficiencies can pose significant challenges for technology adoption in developing economies. These roadblocks can dampen investor confidence, hinder innovation, and impede the growth of technology-based businesses.

Opportunities

1. **Leapfrogging Traditional Development Stages:** Developing economies have the opportunity to leapfrog traditional development stages by adopting advanced technologies. For example, mobile technology has allowed many developing countries to skip the landline telecommunication phase and directly embrace wireless communication. This leapfrogging can accelerate development and bridge the technological gap with developed economies.

2. **Promoting Inclusive Growth:** Technology can enable inclusive growth in developing economies by creating new economic opportunities for marginalized populations. For instance, digital platforms and e-commerce can provide access to markets for small-scale entrepreneurs and farmers, enabling them to earn higher incomes and improve their livelihoods.

3. **Enhancing Efficiency and Productivity:** Technology adoption can lead to increased efficiency and productivity in sectors such as agriculture, manufacturing, and services. Mechanized farming techniques, automation in manufacturing processes, and digital tools for service delivery can help developing economies maximize their resource utilization and boost economic output.

4. **Improving Access to Services:** Technology can enhance access to essential services such as healthcare and education in remote and underserved areas. Telemedicine and e-learning platforms, for instance, can overcome geographical barriers and connect people in remote locations with quality healthcare providers and educational resources.

5. **Fostering Innovation and Entrepreneurship:** Technology can serve as a catalyst for innovation and entrepreneurship in developing economies. It can provide a platform for grassroots innovation and enable the development of technology-based startups. By creating an ecosystem that nurtures entrepreneurship, developing economies can drive economic growth and job creation.

Case Study: Mobile Money in Kenya

A prime example of how technology can transform a developing economy is the success story of mobile money in Kenya. The introduction of the mobile payment system M-Pesa in 2007 revolutionized the financial landscape in the country. M-Pesa enabled individuals to send and receive money, pay bills, and access financial services through their mobile phones, even without a traditional bank account.

The widespread adoption of M-Pesa addressed the challenge of limited access to traditional banking services in Kenya. It provided a secure and convenient avenue for financial transactions, particularly for the unbanked population. The impact of mobile money on the Kenyan economy has been significant, with improved financial inclusion, increased household savings, and enhanced business opportunities.

The success of M-Pesa showcases the transformative power of technology in developing economies. It highlights the opportunities for leapfrogging traditional banking infrastructure and provides valuable lessons in designing and implementing technology solutions to overcome development challenges.

Conclusion

While developing economies face unique challenges in adopting and leveraging technology, they also have significant opportunities for technological advancement. By addressing the challenges of limited access to technology, infrastructural constraints, skills development, financing, and institutional roadblocks, developing economies can harness the power of technology to drive economic growth, promote sustainable development, and improve the lives of their citizens. It requires a collaborative effort between governments, private sector entities, and international organizations to create an enabling environment that supports technology adoption and maximizes its potential for development.

Investigating the role of technology in reshaping global value chains

Introduction

In today's interconnected world, global value chains play a crucial role in shaping the global economy. They involve the sequential stages of production, from the extraction of raw materials to the final product, across multiple countries. However, the advent of technology has significantly reshaped global value chains, leading to increased efficiency, innovation, and interconnectedness. In this section, we will explore how technology has transformed global value chains and the implications for businesses, trade, and economic development.

The Traditional Global Value Chain

Before delving into the role of technology, it is essential to understand the traditional global value chain and its key components. In a traditional global value chain, each stage of production tends to be geographically separated, with specialization occurring in different countries based on comparative advantages. The distribution of labor-intensive tasks, such as manufacturing, occurs in countries with low labor costs, while countries with advanced technology and research capabilities focus on research and development (R&D) and design.

Technology as an Enabler

Technology has become a critical enabler in reshaping global value chains, revolutionizing the way products and services are produced and delivered across borders. Advancements in areas such as information and communication technology (ICT), automation, and digitalization have led to the integration of various stages of production and increased coordination among suppliers, manufacturers, and distributors, regardless of their geographical locations.

Technological Innovations Driving Value Chain Reshaping

Several technological innovations have had a significant impact on reshaping global value chains. Let's explore some of them:

1. **Internet of Things (IoT):** IoT has enabled the interconnectivity of physical devices and systems, allowing real-time data exchange and monitoring of production processes. This technology has improved supply chain visibility, efficiency, and decision-making.

2. **Big Data and Analytics:** The availability of vast amounts of data and advanced analytics tools has facilitated improved forecasting, demand planning, and inventory management. This enables companies to better respond to market dynamics and reduce inefficiencies in the value chain.

3. **Robotics and Automation:** With advancements in robotics and automation, tasks that were previously performed by human workers can now be automated. This has led to increased productivity, cost savings, and higher precision in manufacturing processes.

4. **Cloud Computing:** Cloud computing provides scalable and flexible infrastructure for data storage, collaboration, and software use. It has facilitated the sharing of information and resources across different stages of the value chain, fostering collaboration and streamlining operations.

5. **Artificial Intelligence (AI):** AI technologies, such as machine learning and natural language processing, are being employed in various stages of the value chain, including product design, demand forecasting, quality control, and customer service. AI enables companies to make data-driven decisions, automate repetitive tasks, and improve overall operational efficiency.

Implications for Businesses

The integration of technology into global value chains has significant implications for businesses:

- **Increased Efficiency and Productivity:** Technology has allowed businesses to streamline processes, reduce costs, and improve overall efficiency. Automated systems, real-time data analytics, and predictive modeling enable faster and more informed decision-making, ultimately driving productivity gains.

- **Enhanced Supply Chain Visibility and Resilience:** Technology enables real-time tracking of products and materials throughout the value chain. This visibility enhances supply chain agility, making it easier to identify bottlenecks, optimize inventory management, and mitigate risks.

- **Accelerated Innovation:** Technology enables companies to collaborate and innovate across borders more efficiently. With real-time data sharing and communication tools, businesses can tap into global talent pools, foster partnerships, and enhance their capacity for innovation.

- **Improved Customer Engagement:** Technology allows businesses to engage directly with customers, understand their preferences, and deliver personalized products and services. This contributes to stronger customer relationships and loyalty.

Implications for Trade

The role of technology in reshaping global value chains has profound implications for international trade:

- **Global Market Expansion:** Technology has lowered transaction costs and reduced barriers to entry, enabling small and medium-sized enterprises (SMEs) to participate in global value chains. This has led to increased market access and the expansion of trade opportunities.

- **Disintermediation and New Marketplaces:** Technological advancements, such as e-commerce platforms and online marketplaces, have disrupted traditional intermediary roles in the value chain. Businesses can now interact directly with suppliers and customers, bypassing traditional distribution channels.

- **Changing Trade Patterns:** Technology has facilitated the relocation of certain value chain activities, such as manufacturing, to countries with lower labor costs. This has led to shifts in global trade patterns and increased competition among countries to attract foreign direct investment (FDI) and skilled talent.

- **Intellectual Property Rights Challenges:** The digital nature of value chain activities, such as digital content creation and online transactions, has raised challenges related to intellectual property rights protection and enforcement. Innovations in technology require robust legal frameworks to safeguard intellectual property and promote fair trade practices.

Case Study: Global Value Chains in the Electronics Industry

To further illustrate the role of technology in reshaping global value chains, let's examine the electronics industry, which has undergone significant transformations due to technological advancements.

In the past, the electronics industry relied heavily on vertically integrated value chains, with one company handling all stages of production, from design to manufacturing and distribution. However, with the emergence of technology, this model has significantly changed.

Today, the electronics industry relies on complex and interconnected value chains, with different companies specializing in specific stages of production. For example, companies in South Korea and Taiwan focus on semiconductor design and manufacturing, while China specializes in final assembly and export. This division of labor has allowed for increased efficiency, cost reduction, and specialization within the global electronics value chain.

Furthermore, technology has enabled the emergence of new players in the value chain. For instance, companies like Apple or Samsung no longer manufacture all components in-house. Instead, they outsource the production of specific components to specialized suppliers, who can deliver higher quality and cost-effective components due to economies of scale and expertise. This level of specialization and outsourcing would not have been possible without the integration of technology into global value chains.

Conclusion

Technology has played a transformative role in reshaping global value chains, revolutionizing the way products and services are produced, distributed, and consumed across borders. The integration of technology has brought increased efficiency, productivity, and innovation to businesses, while also influencing trade patterns, market access, and intellectual property rights protection. As technology continues to advance, it is essential for businesses and policymakers to adapt and embrace advancements to remain competitive in the global economy.

Further Reading

1. World Trade Organization. (2017). "The Role of Global Value Chains in Fostering Sustainable Development" - This report provides in-depth analysis and case studies on the role of global value chains in sustainable development, highlighting the influence of technology.

2. McKinsey Global Institute. (2019). "Advanced industries: Reinventing the rules of the game" - This report explores how advanced technologies are reshaping global value chains across different industries.

Exercises

1. Conduct research on a specific industry and examine how technology has reshaped its global value chain. Discuss the implications for businesses and trade in that particular industry.

2. Choose a country and analyze its strategies for attracting and integrating technology-driven global value chains. Assess the successes and challenges faced by the country in this process.

3. Imagine you are a business owner in the traditional value chain model. Develop a strategy to incorporate technology into your value chain to enhance efficiency and competitiveness.

Assessing the impact of digital technologies on cross-border services

The emergence and widespread adoption of digital technologies have revolutionized various aspects of our daily lives, including business practices and the way we conduct cross-border services. In this section, we will explore the impact of digital technologies on cross-border services, examining the benefits, challenges, and potential implications for businesses and economies around the world.

Background

Cross-border services refer to the exchange of services between different countries, either through physical or electronic means. Examples of cross-border services include financial transactions, communication services, professional consulting, and online marketplaces. Traditionally, cross-border services were characterized by complex and time-consuming processes, including physical travel, paperwork, and reliance on traditional communication methods. However, with the advent of digital technologies, such as the internet, cloud computing, and e-commerce platforms, cross-border services have become more efficient, accessible, and cost-effective.

Principles of Digital Technologies

To understand the impact of digital technologies on cross-border services, it is essential to grasp the underlying principles of these technologies. Here are some key principles:

1. Connectivity: Digital technologies enable instant and seamless communication between individuals, businesses, and governments across borders, eliminating the need for physical presence.

2. Speed: With digital technologies, the exchange of information and delivery of services can happen in real-time or near real-time, eliminating delays caused by physical distance.

3. Scalability: Digital technologies have the ability to handle large volumes of data and transactions, allowing businesses to scale up their operations quickly and reach a global audience.

4. Automation: Digital technologies enable the automation of various processes, reducing manual interventions and increasing efficiency in cross-border service delivery.

5. Cost-efficiency: Compared to traditional methods, digital technologies often offer cost savings in terms of transaction fees, travel expenses, and infrastructure requirements.

Benefits of Digital Technologies on Cross-Border Services

The impact of digital technologies on cross-border services has been significant and has brought about several benefits. Let's explore some of these benefits:

1. Increased Access: Digital technologies have lowered barriers to entry for cross-border service providers, allowing businesses of all sizes to offer their services globally. This increased access has created opportunities for small and medium-sized enterprises (SMEs) and startups to compete on a global scale.

2. Enhanced Efficiency: Digital platforms and tools streamline cross-border service delivery by automating processes, eliminating paperwork, and reducing administrative overhead. This increased efficiency translates into faster service delivery and improved customer experiences.

3. Cost Reduction: Digital technologies have reduced costs associated with cross-border services. For instance, international telecommunication costs have significantly decreased, making communication more affordable. Additionally, online marketplaces eliminate the need for physical store presence, reducing rental and operational costs.

4. Expanded Market Reach: Digital technologies enable businesses to reach a global audience without the need for physical presence in multiple countries. This expanded market reach allows businesses to tap into new markets and customer segments, driving revenue growth.

5. Innovation and Collaboration: Digital technologies facilitate innovation and collaboration in the delivery of cross-border services. For example, cloud computing enables real-time collaboration between service providers and customers, fostering creativity and co-creation.

Challenges and Implications

While digital technologies offer numerous benefits to cross-border services, they also present unique challenges and implications. Let's discuss some of these challenges:

1. Data Security and Privacy: Cross-border services often involve the transfer of sensitive data, such as personal information and financial transactions. Digital technologies raise concerns about data security, privacy, and compliance with international regulations. Ensuring the protection of personal data and building trust among customers is crucial for the success of cross-border services.

2. Regulatory and Legal Frameworks: The rapid growth of digital technologies has outpaced the development of regulatory and legal frameworks governing cross-border services. Differences in regulations across countries can create barriers and uncertainties for businesses, hindering their ability to operate seamlessly.

3. Digital Divide: Despite the advancements in digital technologies, there are still disparities in access to technology and internet connectivity, particularly in developing countries. This digital divide can limit the reach and impact of cross-border services, perpetuating economic inequalities.

4. Cybersecurity Threats: As cross-border services heavily rely on digital platforms, they become vulnerable to cybersecurity threats such as data breaches, hacking, and identity theft. Businesses must invest in robust cybersecurity measures to protect their operations and the privacy of their customers.

5. Technological Infrastructure: Digital technologies require a robust technological infrastructure, including reliable internet connectivity and adequate bandwidth. In some regions, the lack of infrastructure can limit the effectiveness of digital cross-border services.

Best Practices and Strategies

To leverage the benefits of digital technologies in cross-border services while addressing the associated challenges, businesses and policymakers can adopt the following best practices and strategies:

1. Collaboration and Partnerships: Businesses can collaborate with local partners and service providers in the target markets to navigate regulatory challenges, gain market insights, and build trust with local customers.

2. Compliance and Data Protection: To address data security and privacy concerns, businesses should comply with international privacy regulations and invest in secure data storage and transmission protocols. Implementing encryption and multi-factor authentication can enhance the security of cross-border services.

3. Technology Infrastructure Investment: Policymakers should prioritize investment in technology infrastructure, including broadband connectivity and data centers, to ensure equitable access to digital technologies for all businesses and individuals.

4. International Standardization: Governments and international organizations can work together to establish harmonized regulatory frameworks, standards, and guidelines for cross-border services. This would simplify compliance requirements and facilitate global trade.

5. Capacity Building and Training: Providing training and capacity-building programs on digital technologies and cross-border service delivery can help

businesses and individuals adapt to the changing landscape. This includes training in cybersecurity best practices, digital marketing strategies, and understanding international regulations.

Case Study: Impact of Digital Platforms on e-commerce

One example of the impact of digital technologies on cross-border services is the rise of e-commerce platforms. Platforms such as Amazon, Alibaba, and eBay have transformed the way businesses sell and consumers purchase goods and services globally. These platforms provide an online marketplace where sellers can reach customers across borders, and buyers can easily access a wide range of products.

The key benefits of digital platforms for cross-border e-commerce include:

1. Global Reach: Digital platforms enable small businesses to access a global customer base, allowing them to expand their reach beyond traditional markets.

2. Consumer Convenience: Digital platforms offer consumers a convenient way to browse and purchase products from different countries, with easy payment and shipping options.

3. Enhanced Trust: Digital platforms build trust among buyers and sellers through buyer reviews, product ratings, and secure payment gateways.

4. Logistics Optimization: Cross-border logistics have become more efficient with the integration of digital platforms, reducing delivery times and costs.

5. Market Intelligence: Digital platforms provide valuable data and insights about customer preferences and market trends, helping businesses make informed decisions and tailor their offerings.

However, challenges such as customs regulations, taxation, and language barriers still exist in cross-border e-commerce. Governments and platform providers are working together to address these challenges and create a more conducive environment for cross-border e-commerce growth.

Conclusion

Digital technologies have brought about a transformative impact on cross-border services, revolutionizing the way businesses operate and interact in a globalized world. While digital technologies offer numerous benefits such as increased efficiency, cost reduction, and expanded market reach, they also present challenges related to data security, regulatory frameworks, and digital divide. By adopting best practices and strategies, businesses and policymakers can harness the full potential of digital technologies to drive inclusive and sustainable growth in cross-border services. The case study on the impact of digital platforms in e-commerce highlights the transformative power of digital technologies in enabling cross-border trade.

Understanding the concept of digital platforms and their role in globalization

In today's interconnected world, digital platforms have become central to the process of globalization. These platforms are online systems or applications that facilitate the exchange of goods, services, information, and ideas between individuals, businesses, and organizations across different countries and regions. They play a crucial role in enabling global communication, collaboration, and commerce in a fast and efficient manner.

Digital platforms can take various forms, including e-commerce platforms, social media networks, search engines, online marketplaces, and sharing economy platforms. These platforms leverage the power of the internet, mobile technologies, and data analytics to connect users and enable interactions on a global scale.

One key characteristic of digital platforms is their ability to connect different actors in what is often referred to as a multi-sided market. For example, an e-commerce platform connects buyers and sellers, a social media platform connects users and advertisers, and a ride-sharing platform connects passengers and drivers. By bringing together different parties, these platforms create a network effect, where the value of the platform increases as more users join and participate.

Digital platforms have several key roles in the process of globalization. Firstly, they enable global trade by connecting buyers and sellers across the world. E-commerce platforms such as Amazon, Alibaba, and eBay have revolutionized the way products are bought and sold, allowing businesses and consumers to reach markets beyond their geographical boundaries. This has opened up new opportunities for small businesses and entrepreneurs to participate in global trade.

Secondly, digital platforms facilitate cross-border collaborations and knowledge sharing. Social media networks like Facebook, Twitter, and LinkedIn allow individuals and organizations to connect, communicate, and collaborate across different countries. These platforms have become important tools for global networking, information sharing, and professional development.

Thirdly, digital platforms play a crucial role in the sharing economy. Platforms like Uber, Airbnb, and TaskRabbit connect individuals who have idle resources or skills with those who need them. This not only creates economic opportunities but also promotes cultural exchange and understanding across borders.

Moreover, digital platforms have significant implications for global value chains. They enable businesses to outsource tasks and processes to specialized providers, leading to the fragmentation of production across different countries. For example, platforms like Upwork and Freelancer connect businesses with freelancers from around the world, allowing them to access a global talent pool and optimize costs.

Digital platforms also generate large amounts of data, which can be used to gain insights and improve decision-making. By analyzing user data and behavior, platforms can personalize services, offer targeted advertising, and enhance the overall user experience. This data-driven approach has the potential to drive innovation, competitiveness, and efficiency in the global marketplace.

However, the rise of digital platforms has also raised concerns and challenges. One of the key challenges is the regulation of these platforms, particularly in areas such as data protection, privacy, competition, and labor rights. Governments and policymakers are grappling with the need to balance innovation and consumer protection in this rapidly evolving landscape.

Furthermore, digital platforms can exacerbate existing inequalities and power imbalances. For example, the dominance of certain platform giants can lead to market concentration and unfair competition. Additionally, the gig economy nature of some platforms can create precarious working conditions and lack of social protection for workers.

In conclusion, digital platforms play a crucial role in the process of globalization by enabling global trade, cross-border collaboration, knowledge sharing, and the sharing economy. They have revolutionized the way businesses operate and individuals connect in today's interconnected world. However, addressing the challenges and ensuring responsible governance of these platforms is essential to harness their full potential for inclusive and sustainable globalization.

Analyzing the challenges of digital divides and the digital economy for developing countries

Digital divides and the unequal access to technology have become major challenges for developing countries in the era of the digital economy. As technology continues to play a central role in driving economic growth and development, the digital divide poses significant barriers to inclusive and equitable participation in the global digital economy. In this section, we will explore the challenges faced by developing countries in bridging the digital divide, and the implications of these challenges for their participation in the digital economy.

Understanding the digital divide

The digital divide refers to the gap between those who have access to digital technologies, such as the internet and smartphones, and those who do not. This divide is often characterized by disparities in access to digital infrastructure, affordability of technology, digital literacy, and the availability of relevant digital content and services. Developing countries often face multiple dimensions of the digital divide, which exacerbate existing economic, social, and educational inequalities.

Disparities in digital infrastructure

One of the major challenges faced by developing countries is the lack of adequate digital infrastructure. This includes limited access to stable and high-speed internet connections, as well as inadequate telecommunications networks and electricity supply. These infrastructure gaps hinder the adoption and use of digital technologies, limiting the potential for economic growth and development. Rural and remote areas within developing countries are particularly affected, as they often lack the necessary infrastructure to connect to the digital world.

Affordability and accessibility

Another significant challenge is the affordability of digital technologies. High costs of devices, such as smartphones and computers, as well as data plans, can make access to technology financially burdensome for many individuals in developing countries. Limited availability and accessibility of internet services further restrict the capabilities of individuals and businesses to participate in the digital economy. This lack of affordability and accessibility creates a barrier to entry and hampers the inclusion of marginalized communities in the digital world.

Digital literacy and skills gap

A critical challenge faced by developing countries is the digital literacy and skills gap. Many individuals lack the necessary skills and knowledge to effectively use digital technologies for their personal and professional needs. This hinders their ability to fully participate in the digital economy and take advantage of the opportunities offered by the digital age. Digital literacy initiatives, such as training programs and educational resources, are essential to bridge this gap and empower individuals to leverage technology for socio-economic advancement.

Limited availability of relevant digital content

In developing countries, the lack of relevant digital content in local languages and contexts poses a significant challenge. Many online platforms, applications, and services primarily cater to developed markets, neglecting the unique needs and preferences of users in developing countries. Additionally, limited local content creation and digital entrepreneurship opportunities further restrict the availability of diverse and culturally relevant digital content. This limitation reduces the attractiveness and usability of digital technologies for individuals and businesses in developing countries.

Implications for participation in the digital economy

The challenges posed by the digital divides have profound implications for the participation of developing countries in the digital economy. Limited access to digital technologies and infrastructure, coupled with affordability and skills barriers, hinder the integration of individuals and businesses into global digital value chains. This limits the potential for job creation, innovation, and economic growth. Developing countries are at risk of being left behind in the digital revolution, further exacerbating existing inequalities and hindering their overall development prospects.

To address these challenges, developing countries need comprehensive strategies that prioritize bridging the digital divide. These strategies should encompass various dimensions, including expanding digital infrastructure, promoting affordable access, enhancing digital literacy programs, fostering local content creation, and empowering digital entrepreneurship. Collaboration between governments, private sector entities, and civil society organizations is crucial for implementing these strategies effectively.

By bridging the digital divide and ensuring equitable access to technology, developing countries can unlock the transformative potential of the digital economy. This can lead to increased productivity, improved social inclusion, and sustainable development. However, it is essential to address the challenges and barriers faced by developing countries to create an inclusive and digitally-enabled future for all.

Resources

1. World Bank - "World Development Report 2016: Digital Dividends" 2. United Nations Conference on Trade and Development (UNCTAD) - "Information Economy Report 2019: Digitalization, Trade, and Development" 3. International

Telecommunication Union (ITU) - "Measuring digital development: Facts and figures 2020" 4. Global System for Mobile Communications Association (GSMA) - "The Mobile Economy 2020" 5. Deloitte - "Digital Africa: Powering the Economy"

Discussing the Role of International Organizations in Promoting Technology-Driven Globalization

In today's interconnected world, technology plays a vital role in driving globalization. International organizations have a significant responsibility in promoting and facilitating the integration of technology into the global economy. These organizations facilitate cooperation between countries, provide technical assistance, support research and development initiatives, and establish policies and frameworks to ensure a fair and sustainable technology-driven globalization. In this section, we will discuss the role of international organizations in promoting technology-driven globalization and explore their key initiatives and contributions.

The Role of International Organizations in Facilitating Technology Diffusion

International organizations, such as the World Bank, International Monetary Fund (IMF), World Trade Organization (WTO), United Nations (UN), Organization for Economic Co-operation and Development (OECD), and regional bodies like the European Union (EU), play a crucial role in facilitating the diffusion of technology across borders. They promote technology transfer, knowledge sharing, and capacity building initiatives to bridge the digital divide between developed and developing economies.

These organizations facilitate the exchange of best practices, policies, and regulations to create an enabling environment for technology adoption and innovation. They also provide technical assistance and financing to help countries build the necessary technological infrastructure and capabilities, especially in developing regions. For instance, the World Bank's Global Information and Communication Technology (ICT) Department supports ICT infrastructure development, digital skills training, and e-government initiatives in developing countries.

Promoting Access to Technology and Digital Inclusion

One of the key concerns in technology-driven globalization is the digital divide, which refers to the gap between those who have access to digital technologies and those who do not. International organizations work towards promoting digital inclusion and ensuring equal access to technology for all countries, regardless of their economic status.

These organizations collaborate with governments, the private sector, and civil society to develop policies and initiatives that aim to bridge the digital divide. They advocate for affordable and reliable internet connectivity, technology infrastructure, and digital literacy programs. For example, the UN's Sustainable Development Goals (SDGs) include targets for universal and affordable internet access and digital skills development.

Establishing Standards and Guidelines

International organizations play a critical role in establishing standards and guidelines for technology and innovation. These standards ensure interoperability, compatibility, and safety of technologies, facilitating their adoption and use across different countries and industries.

For instance, the International Electrotechnical Commission (IEC) develops international standards for electrical and electronic technologies, ensuring their compatibility and safety. The International Organization for Standardization (ISO) develops standards for various fields, including information technology and telecommunications, to enable interoperability and enhance technological cooperation between countries.

Furthermore, international organizations promote the protection of intellectual property rights (IPR) to encourage innovation and technology transfer. They work towards establishing fair and balanced IPR regimes that safeguard the interests of both innovators and society at large.

Facilitating International Cooperation and Collaboration

International organizations facilitate international collaboration and cooperation in technology-driven globalization. They provide platforms for dialogue, knowledge sharing, and joint initiatives between countries, academia, industry, and research institutions.

For example, the OECD's Digital Economy Outlook report provides analyses and policy recommendations to promote digital transformation and cooperation among member countries. The EU's Horizon 2020 program encourages collaborative research and innovation projects among European countries and beyond.

International organizations also promote partnerships between the public and private sectors to leverage resources and expertise for technology-driven development. They facilitate public-private partnerships (PPPs) to support technology transfer, innovation, and capacity building initiatives.

Addressing Ethical and Societal Impacts

As technology-driven globalization progresses, international organizations recognize the need to address the ethical, social, and cultural impacts of technology adoption. They develop guidelines and policies to ensure that technology is harnessed responsibly and for the benefit of all.

For instance, the United Nations Educational, Scientific and Cultural Organization (UNESCO) advocates for the ethical use of artificial intelligence (AI) and promotes principles such as transparency, accountability, and inclusivity in AI development and deployment.

International organizations also address the social implications of technology-driven globalization, such as job displacement and inequality. They work towards ensuring that the benefits of technology are distributed equitably and that no one is left behind in the digital age. These organizations support education and skills development programs to equip individuals with the necessary abilities to adapt to technological changes.

Promoting Sustainable and Responsible Technology-driven Development

Sustainability is a key concern in technology-driven globalization, as technological advancements can have both positive and negative environmental impacts. International organizations promote sustainable development by encouraging the adoption of clean technologies, promoting energy efficiency, and advocating for environmentally friendly practices.

For example, the United Nations Framework Convention on Climate Change (UNFCCC) supports the development and transfer of climate-friendly technologies to help countries mitigate and adapt to climate change. The International Renewable Energy Agency (IREA) promotes the adoption of renewable energy technologies worldwide.

International organizations also address the responsible development and use of emerging technologies, such as nanotechnology and biotechnology. They establish frameworks and guidelines to ensure the ethical and safe deployment of these technologies, balancing innovation with potential risks.

Conclusion

International organizations play a vital role in promoting technology-driven globalization. They facilitate technology diffusion, promote digital inclusion, establish standards and guidelines, foster international cooperation, address ethical and societal impacts, and promote sustainable and responsible technology-driven development. Their efforts contribute to the creation of a globally connected and technologically advanced world that benefits all nations and societies. By leveraging the expertise and resources of these organizations, countries can navigate the opportunities and challenges of technology-driven globalization in a collaborative and responsible manner.

Examining the Ethical Implications of Technology-Driven Globalization

Technology-driven globalization has brought about numerous benefits and opportunities for societies worldwide. However, it is equally important to critically analyze and understand the ethical implications stemming from this rapid integration of technology into our global economy. In this section, we will explore some of the key ethical concerns associated with technology-driven globalization and highlight strategies for addressing them.

Privacy and Data Security

One of the foremost ethical issues arising from technology-driven globalization is the preservation of privacy and data security. As digital technologies continue to advance, the collection, storage, and utilization of personal data have become increasingly prevalent. While this data can be beneficial for various purposes such as targeted advertising and improved service delivery, it also raises concerns related to unauthorized access, misuse, and potential violations of individual privacy.

To address these ethical concerns, it is crucial to establish comprehensive legal frameworks and regulations governing the collection and use of personal data. Governments and international organizations must work together to develop robust privacy policies that strike a balance between facilitating innovation and safeguarding individuals' rights. Additionally, the integration of strong encryption mechanisms and cybersecurity protocols can ensure data security and protect against unauthorized access.

Labor Rights and Working Conditions

Globalization, facilitated by technology, has enabled businesses to expand their operations across borders, often resulting in the outsourcing of labor to countries with lower costs. While this presents economic opportunities, it also raises ethical concerns regarding labor rights and working conditions in these regions.

Ensuring fair treatment and decent working conditions for all individuals involved in the global supply chain becomes imperative. Companies must adopt ethical labor practices, providing fair wages, safe working environments, and reasonable working hours. Governments and international bodies can play a vital role by setting and enforcing labor standards, conducting regular audits, and providing support to workers' rights organizations.

Environmental Impact

The acceleration of global trade, enabled by technology, has led to unprecedented levels of resource consumption and environmental degradation. As technology-driven globalization continues to increase economic activities, it is crucial to address the ethical implications related to sustainability and environmental impact.

To address these concerns, businesses and governments must adopt sustainable practices. This includes investing in renewable energy sources, promoting circular economy models, and reducing carbon emissions. Additionally, international agreements and collaborations can facilitate the exchange of knowledge and best practices to mitigate the environmental impact of technology-driven globalization.

Equity and Access

While technology-driven globalization has the potential to bridge the gap between developed and developing countries, it also carries the risk of exacerbating existing inequalities. Unequal access to technology and limited digital literacy can create a "digital divide" that further marginalizes vulnerable populations.

To ensure equitable access, governments and organizations must invest in digital infrastructure and promote digital literacy programs. Additionally, providing affordable and accessible internet connectivity can enable individuals in underserved areas to participate in the benefits of technology-driven globalization. Collaborative efforts between governments, civil society, and the private sector are crucial for achieving greater equity in access to technology.

Cultural Diversity and Digital Divide

The widespread adoption of technology can lead to the erosion of cultural diversity and traditional knowledge. As globalization spreads, local cultures and languages may face the risk of being overshadowed by dominant global trends.

To preserve cultural diversity, governments should implement policies that promote local content, languages, and traditional knowledge systems in the digital sphere. Additionally, fostering an inclusive digital environment that allows for diverse voices and perspectives can help mitigate the risks associated with the digital divide.

Conclusion

Technology-driven globalization brings significant economic and social benefits. However, it also carries ethical implications that must be addressed to ensure sustainable and inclusive development. By considering the ethical dimensions of privacy, labor rights, environmental impact, equity, access, and cultural diversity, we can shape technology-driven globalization to promote a more equitable and responsible global economy. It is crucial for stakeholders, including governments, businesses, and individuals, to work together in addressing these ethical concerns to create a future that benefits all.

Exploring strategies for promoting fair and sustainable technology-driven globalization

In today's interconnected world, technology plays a crucial role in driving globalization and shaping economies. As countries become increasingly reliant on technology, it is essential to explore strategies that promote fair and sustainable technology-driven globalization. This section will discuss various approaches and initiatives that can be undertaken to ensure that globalization through technology benefits all stakeholders and contributes to sustainable development.

Promoting Digital Inclusion

One of the key strategies for promoting fair and sustainable technology-driven globalization is to focus on digital inclusion. This involves ensuring that all individuals and communities have equal access to digital technologies and the opportunities they offer. The digital divide, which refers to the gap between those who have access to digital technologies and those who do not, is a significant barrier to fair globalization.

To address this issue, governments and organizations can implement policies and programs that aim to bridge the digital divide. This can include infrastructure development initiatives to expand internet access to underserved areas, providing affordable devices, and offering digital literacy training programs. Additionally, initiatives that promote the localization of technology, such as developing locally relevant content and applications, can also contribute to digital inclusion.

Strengthening Data Protection and Privacy

In the era of globalization, data has become a valuable asset, driving economic growth and innovation. However, concerns around data protection and privacy have also emerged. To ensure fair and sustainable technology-driven globalization, it is crucial to strengthen data protection and privacy regulations.

Governments can enact comprehensive data protection laws that safeguard individuals' rights and ensure responsible data handling by businesses and organizations. These regulations should establish clear guidelines for data collection, storage, and usage, emphasizing consent and transparency. Additionally, international cooperation on data protection frameworks can help in promoting fair globalization by establishing common standards and facilitating cross-border data transfers while protecting individual privacy.

Fostering Collaboration and Knowledge Sharing

Promoting fair and sustainable technology-driven globalization requires fostering collaboration and knowledge sharing among nations, organizations, and individuals. Open innovation approaches, where knowledge and ideas are shared openly, can drive technological advancements and benefit a larger audience.

International collaboration platforms can be established to facilitate the exchange of technology-related knowledge and expertise. These platforms can bring together researchers, entrepreneurs, policymakers, and other stakeholders, fostering collaboration on research and development, capacity building, and technology transfer. By encouraging the exchange of best practices and lessons learned, such initiatives can accelerate global technological progress while ensuring that the benefits are shared across nations.

Promoting Ethical Technological Standards

As technology becomes increasingly ingrained in our lives, it is crucial to promote ethical technological standards that align with sustainable development goals. This involves considering social, environmental, and economic impacts throughout the technology lifecycle, from design and development to disposal.

Governments and industry bodies can establish regulations and standards that encourage the adoption of ethical practices in technology development and use. This can include guidelines for responsible innovation, promoting diversity and inclusivity in the technology sector, and encouraging the development of technologies that address pressing global challenges, such as climate change and inequality.

Addressing Digital Monopolies and Power Imbalances

In the context of technology-driven globalization, addressing digital monopolies and power imbalances is essential to ensure fair competition and a level playing field for all stakeholders. The concentration of power in the hands of a few tech giants can hinder fair globalization and limit opportunities for smaller players.

Governments can take measures to prevent and address monopolistic practices in the technology sector through robust competition policies and antitrust regulations. Additionally, initiatives that promote open and interoperable standards can help create an ecosystem that allows for fair competition and innovation. By promoting a diverse and inclusive technology sector, it becomes possible to drive fair and sustainable globalization.

Educating and Developing Skilled Workforce

For technology-driven globalization to be fair and sustainable, it is essential to focus on education and skills development. Upskilling and reskilling of the workforce are critical to ensure that individuals can participate in and benefit from the opportunities created by technology.

Governments and educational institutions can collaborate to develop comprehensive programs that equip individuals with the necessary digital skills and competencies. These programs can include technical training, digital literacy initiatives, and entrepreneurship education. Additionally, investing in research and development in emerging technologies can help countries stay competitive and contribute to fair globalization.

Promoting Environmental Sustainability

Sustainable development is intrinsically linked to environmental sustainability. Globalization through technology should be pursued in a manner that minimizes negative environmental impacts and maximizes positive contributions.

Fostering the development and adoption of clean technologies can contribute to sustainable technology-driven globalization. Governments can provide incentives for the development and use of renewable energy technologies, promote energy-efficient practices, and support green innovation. Embracing circular economy principles, such as recycling and waste reduction, within the technology sector can also contribute to sustainability goals.

Ensuring Ethical Supply Chains

Promoting fair and sustainable technology-driven globalization requires ensuring ethical supply chains for technology products. This involves considering the social and environmental impacts of the entire supply chain, from raw material extraction to product manufacturing and disposal.

Governments and organizations can establish standards and certification programs that promote ethical practices within the technology supply chain. This can include addressing issues such as child labor, forced labor, and environmental pollution. By promoting responsible sourcing and production, stakeholders can ensure that technology-driven globalization adheres to sustainable development principles.

Monitoring and Evaluation

Finally, to ensure the effectiveness of strategies for promoting fair and sustainable technology-driven globalization, robust monitoring and evaluation mechanisms should be in place. This involves tracking progress towards goals and assessing the impact of initiatives.

Governments and international organizations can establish frameworks for monitoring and evaluating the outcomes of technology-driven globalization initiatives. This can include indicators such as digital inclusion rates, environmental sustainability metrics, and measures of social equity and inclusion. Regular evaluations can help identify areas for improvement and guide the refinement of strategies to ensure fairness and sustainability in technology-driven globalization.

In conclusion, promoting fair and sustainable technology-driven globalization requires a comprehensive and multi-faceted approach. By focusing on digital inclusion, data protection, collaboration, ethical standards, competition, education, environmental sustainability, ethical supply chains, and monitoring, stakeholders can ensure that technology-driven globalization benefits all and contributes to sustainable development. It is through these strategies that we can shape a future where technology serves as a force for positive change on a global scale.

Technology and Sustainable Development

Understanding the role of technology in sustainable development

Examining the concept of sustainable development and its relationship with technology

Sustainable development is a concept that has gained significant attention in recent years, as societies strive to find ways to meet current needs without compromising the ability of future generations to meet their own needs. It involves the integration of economic, social, and environmental considerations in decision-making processes to ensure long-term well-being for both people and the planet.

At its core, sustainable development aims to find a balance between economic growth, social progress, and environmental protection. This requires a holistic approach that takes into account not only economic factors, but also social equity, cultural diversity, and environmental sustainability. Technology plays a crucial role in achieving sustainable development goals by providing solutions to the complex challenges we face.

One of the key aspects of sustainable development is the promotion of environmentally friendly practices. Technology has a significant impact on the environment, both positive and negative. On the positive side, technological advancements have led to the development of clean technologies that help reduce pollution, conserve resources, and mitigate climate change. For example, the transition from fossil fuels to renewable energy sources such as solar, wind, and hydroelectric power has become possible due to advancements in technology. These clean technologies not only help in reducing greenhouse gas emissions but also promote the efficient use of resources.

On the other hand, technology can also contribute to environmental degradation if not used responsibly. The production and disposal of electronic devices, for instance, can lead to hazardous waste and resource depletion. Therefore, sustainable development requires the adoption of technologies that minimize environmental impacts throughout their lifecycle, from design to disposal. This includes promoting eco-design principles, recycling programs, and the use of environmentally friendly materials.

Sustainable development also seeks to address social and economic inequities. Technology can play a vital role in promoting social inclusion and reducing inequalities by providing access to information, education, and opportunities. For example, the Internet and digital technologies have revolutionized access to information and communication, empowering individuals and communities to participate in decision-making processes and access essential services.

Moreover, technology can contribute to economic growth and poverty alleviation by creating new job opportunities and improving productivity. By enabling innovation and promoting entrepreneurship, technology can help develop new industries and create sustainable livelihoods. For example, the rise of the sharing economy, driven by technology platforms, has created alternative sources of income for many individuals.

However, it is important to recognize that technology alone cannot solve all the challenges of sustainable development. It needs to be accompanied by effective policies, regulations, and institutional frameworks that support its deployment and ensure its benefits are shared by all. Furthermore, public engagement and participation are essential to ensure that technological advancements align with the values, needs, and aspirations of society.

In summary, sustainable development requires a careful consideration of economic, social, and environmental factors. Technology plays a crucial role in achieving sustainable development goals by providing solutions to the complex challenges we face. It has the potential to promote environmental sustainability, address social inequalities, and drive economic growth. However, it is important to ensure that technology is used responsibly and in line with the principles of sustainable development.

Analyzing the potential of clean technologies for addressing environmental challenges

In recent years, there has been an increasing recognition of the urgent need to address environmental challenges such as climate change, air pollution, and resource depletion. Clean technologies have emerged as a promising solution to mitigate these challenges and promote sustainable development. In this section, we will analyze the potential of clean technologies in addressing environmental issues and discuss their application in various sectors.

Understanding clean technologies

Clean technologies, also known as green technologies or environmental technologies, refer to the use of innovative processes, products, and services that minimize negative environmental impact and promote the efficient use of resources. These technologies aim to reduce pollution, conserve energy and water, and support the transition to a low-carbon economy.

Clean technologies cover a broad range of sectors, including energy, transportation, waste management, water treatment, and agriculture. They encompass a wide variety of innovations, such as renewable energy systems, energy-efficient appliances, emission control technologies, sustainable farming practices, and recycling processes.

The adoption of clean technologies not only helps mitigate environmental challenges but also presents new economic opportunities, such as job creation, investment attraction, and export potential. As such, the development and deployment of clean technologies have become an essential element of sustainable development strategies worldwide.

Addressing climate change

One of the most pressing environmental challenges facing our planet is climate change. The burning of fossil fuels for energy generation and transportation has resulted in the accumulation of greenhouse gases in the atmosphere, leading to global warming. Clean technologies play a crucial role in addressing climate change by reducing greenhouse gas emissions and promoting the use of renewable energy sources.

Renewable energy technologies, such as solar power, wind power, and hydropower, offer sustainable alternatives to fossil fuel-based energy generation. These technologies harness natural resources without depleting them and produce electricity with significantly lower or zero emissions of greenhouse gases. For instance, solar photovoltaic systems convert sunlight directly into electricity, while wind turbines generate electricity from the kinetic energy of the wind. By deploying these technologies on a large scale, the dependency on fossil fuels can be reduced, resulting in a significant reduction in carbon emissions.

Additionally, clean technologies enable energy efficiency improvements in various sectors, such as buildings, manufacturing, and transportation. Energy-efficient technologies, such as LED lighting, advanced insulation materials, and hybrid/electric vehicles, help reduce energy consumption and decrease greenhouse gas emissions. These technologies not only contribute to climate change mitigation but also offer economic benefits through energy cost savings and improved productivity.

Improving air quality

Air pollution, particularly in urban areas, has detrimental effects on human health and the environment. Clean technologies offer effective solutions for improving air quality by reducing emissions of harmful pollutants from industrial processes, transportation, and energy production.

In the transportation sector, clean technologies such as electric vehicles (EVs) and hybrid vehicles significantly reduce tailpipe emissions of pollutants, including nitrogen oxides (NOx), particulate matter (PM), and carbon monoxide (CO). Additionally, advancements in emission control technologies, such as catalytic converters, help minimize the release of harmful pollutants from internal combustion engines.

In industries, the adoption of clean technologies, such as advanced emissions control systems, pollutant capture and treatment technologies, and cleaner production processes, can reduce the emission of air pollutants. These technologies utilize advanced filtration, scrubbing, and catalytic processes to remove pollutants before they are released into the atmosphere.

Furthermore, clean technologies also play a crucial role in the energy sector in reducing air pollution. The transition from fossil fuel-based power generation to renewable energy sources significantly reduces the emission of air pollutants, such as sulfur dioxide (SO2), nitrogen oxides, and fine particulate matter. The use of clean technologies, such as solar and wind power plants, as a sustainable alternative to coal-fired power plants, helps improve air quality and protect human health.

Promoting sustainable agriculture

Agriculture is a significant contributor to environmental challenges, including deforestation, water pollution, and greenhouse gas emissions. Clean technologies offer innovative solutions to promote sustainable agricultural practices and address these environmental issues.

Precision agriculture technologies, such as remote sensing, geographic information systems (GIS), and global positioning systems (GPS), enable farmers to optimize resource use and reduce environmental impacts. By collecting data on soil conditions, crop growth patterns, and weather forecasts, farmers can make informed decisions regarding the application of fertilizers, irrigation, and pest control. This helps minimize the use of chemical inputs, reduce water consumption, and maximize crop productivity while minimizing the negative environmental consequences.

Additionally, clean technologies in agriculture encompass practices such as organic farming, agroforestry, and sustainable aquaculture. These approaches prioritize the conservation of natural resources, biodiversity, and soil health while minimizing the use of synthetic fertilizers and pesticides. For instance, organic farming methods promote the use of natural fertilizers, crop rotation, and biological pest control to enhance soil fertility and prevent water pollution.

Moreover, the use of clean technologies in agriculture also includes the development of efficient irrigation systems, which help optimize water use and minimize water wastage. Drip irrigation systems, for example, deliver water directly to plant roots, reducing water loss due to evaporation and runoff.

Challenges and opportunities

While clean technologies offer significant potential in addressing environmental challenges, several barriers and challenges hinder their widespread adoption.

Cost is often a significant barrier to the deployment of clean technologies. Many clean technologies, such as renewable energy systems and energy-efficient appliances, require high upfront investments, which may deter potential adopters. However, as technology advances and economies of scale are achieved, the costs associated with clean technologies are gradually decreasing, making them more accessible and economically viable.

Another challenge is the lack of supportive policies and regulatory frameworks. Governments need to implement effective incentive mechanisms, such as feed-in tariffs and tax credits, to encourage the adoption of clean technologies. Additionally, policies promoting research and development, technology transfer, and capacity building can accelerate the innovation and diffusion of clean technologies.

Furthermore, the integration of clean technologies into existing infrastructure and systems presents technical and logistical challenges. For instance, the incorporation of renewable energy sources into the electrical grid requires careful planning and investment in grid infrastructure and storage technologies.

Despite these challenges, the global shift towards clean technologies presents significant opportunities for economic growth, job creation, and sustainable development. The clean technology sector has become a major source of innovation, entrepreneurship, and investment, attracting funds and talent from around the world.

In conclusion, clean technologies have immense potential in addressing environmental challenges and promoting sustainable development. Through the deployment of clean technologies in sectors such as energy, transportation, waste management, water treatment, and agriculture, we can reduce greenhouse gas emissions, improve air quality, and promote resource efficiency. However, overcoming barriers to adoption and fostering supportive policy environments are crucial for unlocking the full potential of clean technologies in addressing environmental challenges and achieving a sustainable future.

Understanding the importance of energy-efficient technologies for reducing carbon emissions

In today's world, the issue of climate change and its adverse effects on the environment has become a significant concern. One of the leading contributors to climate change is the emission of carbon dioxide (CO_2) and other greenhouse gases into the atmosphere. These emissions primarily result from the burning of fossil fuels for energy consumption. In order to mitigate the impact of climate change, it is crucial to reduce carbon emissions and transition towards more sustainable energy sources. Energy-efficient technologies play a vital role in achieving this goal by minimizing energy waste and reducing the overall carbon footprint of various sectors.

The importance of energy efficiency

Energy efficiency refers to the ability to accomplish a desired task or output with the least amount of energy input. It involves using energy-efficient technologies and practices to optimize energy consumption. The importance of energy efficiency in reducing carbon emissions can be understood through the following points:

1. **Economic benefits:** Energy efficiency measures help to lower energy consumption, thereby reducing energy costs for households, businesses, and industries. By making better use of energy resources, energy-efficient technologies can lead to significant financial savings in the long run.

2. **Environmental benefits:** The reduction of carbon emissions through energy efficiency helps to mitigate climate change and its adverse effects on the environment. By consuming less energy, there is a reduced demand for fossil fuels, resulting in lower greenhouse gas emissions and air pollution.

3. **Energy security:** Energy efficiency reduces dependence on imported fossil fuels, as it allows for the optimization of existing energy resources. By reducing energy waste and improving energy productivity, countries can enhance their energy security and reduce vulnerability to external supply disruptions.

4. **Job creation:** The adoption of energy-efficient technologies creates employment opportunities, especially in industries related to renewable energy, energy management, and building retrofitting. This helps to stimulate economic growth and promote sustainable development.

Energy-efficient technologies

Energy-efficient technologies encompass a wide range of solutions and practices that aim to maximize energy savings in different sectors. These technologies can be applied in buildings, transportation, industry, and the power sector. Some key energy-efficient technologies include:

1. **Building insulation and energy management systems:** By improving insulation, using energy-efficient windows, and implementing smart energy management systems, buildings can reduce heating and cooling energy requirements.

2. **LED lighting:** Light-emitting diode (LED) lighting consumes significantly less energy compared to traditional incandescent or fluorescent lighting. LED technology has advanced rapidly in recent years, providing improved quality of light and longer lifespans.

3. **Energy-efficient appliances and equipment:** Energy Star-rated appliances and high-efficiency equipment, such as air conditioners, refrigerators, and heating systems, can result in significant energy savings without sacrificing performance.

4. **Fuel-efficient vehicles:** The transportation sector is a major contributor to carbon emissions. Energy-efficient vehicles, including hybrid and electric vehicles, offer reduced fuel consumption and lower emissions, helping to mitigate climate change.

5. **Industrial process optimization:** Industries can improve energy efficiency by adopting process optimization techniques, such as cogeneration, waste heat recovery, and efficient control systems.

6. **Renewable energy technologies:** The integration of renewable energy sources, such as solar power, wind power, and hydropower, into the energy mix helps to reduce reliance on fossil fuels and lower carbon emissions.

7. **Smart grids and energy storage:** Smart grid systems enable efficient management of electricity distribution, reducing losses and optimizing energy usage. Energy storage technologies improve the reliability and flexibility of renewable energy systems, allowing for better integration into the grid.

Policy and challenges

While the benefits of energy-efficient technologies are clear, several challenges need to be addressed to facilitate their widespread adoption:

1. **Lack of awareness and knowledge:** Many individuals, businesses, and governments are not fully aware of the benefits and potential of energy-efficient technologies. Raising awareness and providing education and training programs are essential to overcome this barrier.

2. **Initial costs and financial barriers:** The upfront costs of adopting energy-efficient technologies can be a hurdle for many individuals and businesses. Governments can play a crucial role in providing incentives, grants, and low-interest loans to encourage investment in energy-efficient solutions.

3. **Market barriers and lack of standards:** In some cases, the market for energy-efficient technologies may be underdeveloped. Governments can promote the development of standards, certifications, and labeling schemes to ensure the quality and performance of energy-efficient products.

4. **Infrastructure limitations:** The integration of renewable energy sources and the deployment of energy-efficient technologies may require upgrades to existing infrastructure. Governments and energy providers need to invest in grid modernization and energy infrastructure development.

5. **Policy and regulatory frameworks:** Governments should establish supportive policy frameworks, including energy efficiency standards, carbon pricing mechanisms, and renewable energy targets, to create an enabling environment for the adoption of energy-efficient technologies.

6. **International cooperation:** Collaboration among countries is crucial to address global climate challenges and promote the adoption of energy-efficient technologies worldwide. Sharing best practices, technological advancements, and financial resources can accelerate the transition to a low-carbon economy.

Case study: Energy-efficient buildings

Buildings account for a significant portion of global energy consumption and carbon emissions. Energy-efficient buildings aim to reduce energy demand for heating, cooling, and lighting while ensuring comfort and productivity. The following measures can be implemented to improve the energy efficiency of buildings:

- **Building envelope improvements:** Enhancing insulation, using high-performance windows, and minimizing air leakage can reduce heat transfer and energy loss.

- **Efficient heating, ventilation, and air conditioning (HVAC) systems:** Installing energy-efficient HVAC systems and optimizing their operation through advanced controls can significantly reduce energy consumption.

- **Lighting optimization:** Utilizing natural lighting, installing energy-efficient lighting fixtures, and implementing lighting controls can lead to substantial energy savings.

- **Energy management systems:** Implementing smart building automation systems can optimize energy usage, monitor energy consumption, and identify areas for improvement.

- **Renewable energy integration:** Incorporating renewable energy systems, such as solar panels or geothermal heat pumps, can further reduce reliance on fossil fuels and lower carbon emissions.

For example, the Empire State Building in New York City underwent a comprehensive energy retrofit, which resulted in a 38% reduction in energy consumption and annual energy savings of $4.4 million. The retrofit included upgrading lighting systems, improving insulation, and optimizing HVAC systems.

Conclusion

Energy-efficient technologies play a critical role in reducing carbon emissions and mitigating climate change. By adopting energy-efficient practices and technologies in various sectors, we can achieve substantial energy savings, lower costs, and enhance environmental sustainability. However, overcoming barriers to adoption and fostering innovation and collaboration are essential to realizing the full potential of energy-efficient technologies. Governments, businesses, and individuals all have a role to play in creating a sustainable and energy-efficient future for generations to come.

Discussing the impact of technology on sustainable urbanization and transportation

Urbanization and transportation are critical components of economic and social development. As the world becomes increasingly urbanized, with more people living in cities, the need for sustainable urbanization and transportation solutions has become urgent. Technology plays a crucial role in addressing these challenges and shaping cities of the future. In this section, we will explore the impact of technology on sustainable urbanization and transportation, discussing key principles, challenges, and opportunities.

Principles of Sustainable Urbanization

Sustainable urbanization aims to create cities that are environmentally, socially, and economically sustainable. It involves the planning, design, and management of urban areas with a focus on long-term viability and resilience. Several principles guide sustainable urbanization:

1. **Compact Development:** Promoting higher population densities and mixed land-use patterns to reduce urban sprawl and promote efficient land use.

2. **Transit-Oriented Development (TOD):** Designing urban areas around public transportation systems to reduce car dependency and encourage walking, cycling, and the use of public transit.

3. **Energy Efficiency:** Emphasizing energy-efficient buildings, sustainable infrastructure, and clean energy sources to reduce greenhouse gas emissions and mitigate climate change.

4. **Green Spaces and Biodiversity:** Preserving and enhancing green spaces, parks, and natural habitats within urban areas to improve air quality, promote biodiversity, and enhance the quality of life.

5. **Social Inclusivity:** Ensuring equitable access to housing, transportation, education, healthcare, and other essential services for all residents, regardless of income or social status.

6. **Resilience and Disaster Preparedness:** Incorporating climate adaptation and disaster risk reduction measures to enhance urban resilience and reduce vulnerability to natural hazards.

These principles guide the development of sustainable urban areas and provide a framework for leveraging technology to achieve these goals.

Role of Technology in Sustainable Urbanization

Technology has the potential to revolutionize sustainable urbanization by improving efficiency, resilience, and livability. Here are some key areas in which technology is making a significant impact:

1. **Smart Grids and Energy Management Systems:** Advanced energy management systems and smart grids enable cities to optimize the generation, distribution, and consumption of electricity, reducing energy waste and promoting the integration of renewable energy sources.

2. **Intelligent Transportation Systems (ITS):** ITS uses technology and communication networks to enhance transportation efficiency, reduce congestion, and improve safety. Examples include intelligent traffic management systems, real-time transit information, and the use of sensors for parking management.

3. **Smart Buildings and Infrastructure:** Incorporating smart technologies into buildings and infrastructure improves energy efficiency, enhances resource management, and allows for real-time monitoring and control. Features such as automated lighting, smart HVAC systems, and water management systems contribute to sustainable urban development.

4. **Sustainable Mobility Solutions:** Technology-enabled solutions, such as ride-sharing services, bike-sharing systems, and electric vehicle charging infrastructure, promote sustainable modes of transportation and reduce reliance on private vehicles.

5. **Urban Sensing and Big Data Analytics:** The deployment of sensors and data analytics platforms enables real-time monitoring of various urban parameters, such as air quality, noise levels, and waste management. This data can inform evidence-based decision-making and improve urban planning and resource allocation.

6. **Smart City Platforms:** Integrated smart city platforms bring together various technologies and data sources to enable efficient city management. These platforms facilitate the delivery of public services, support citizen engagement, and provide a foundation for innovative urban solutions.

These technologies not only contribute to sustainable urbanization but also present significant economic opportunities, driving innovation, job creation, and economic growth.

Challenges and Opportunities

While technology has immense potential for sustainable urbanization, it also presents several challenges that need to be addressed:

- **Cost and Affordability:** Many sustainable technologies, such as renewable energy systems and smart infrastructure, can be costly to implement and may face affordability issues, particularly in developing economies. Addressing these cost barriers is essential for widespread adoption.

- **Digital Divide:** Technological advancements can exacerbate existing social inequalities if access to technology and digital literacy is not equitable. Efforts should be made to bridge the digital divide and ensure that everyone can benefit from technology-enabled solutions.

- **Data Management and Privacy:** The use of sensors and data collection in smart cities raises concerns about data management, privacy, and security. Clear policies and regulations are needed to address these concerns and protect the rights of individuals.

- **Infrastructure Integration:** Implementing technology-enabled solutions requires integrating various systems and infrastructure components. Coordination among different stakeholders and sectors is crucial to ensure seamless integration and interoperability.

Despite these challenges, technology-driven sustainable urbanization offers numerous opportunities:

- **Enhanced Resource Efficiency:** Technology enables optimal use of resources, reducing energy consumption, waste generation, and water usage in urban areas.

- **Improved Quality of Life:** Smart technologies enhance the livability and well-being of urban residents by providing efficient transportation, better access to services, and a healthier environment.

- **Economic Growth and Job Creation:** The deployment of sustainable urban technologies stimulates economic growth and creates employment opportunities in sectors such as clean energy, information technology, and urban planning.

- Innovation and Collaboration: Technology-driven urbanization fosters innovation and collaboration among stakeholders, including governments, businesses, academia, and citizens, leading to creative solutions and new partnerships.

- Climate Change Mitigation and Resilience: By reducing greenhouse gas emissions, optimizing resource usage, and incorporating climate adaptation measures, technology improves urban resilience and contributes to global efforts in combating climate change.

Case Study: Barcelona's Smart City Initiatives

One notable example of technology-driven sustainable urbanization is Barcelona's smart city initiatives. Through comprehensive planning and the integration of various technologies, Barcelona has transformed itself into a leading smart city.

Smart Grid and Energy Efficiency: Barcelona has implemented a smart grid system that optimizes energy distribution and consumption, reducing energy waste and emissions. Smart meters enable residents to monitor their energy usage in real-time, encouraging energy conservation.

Smart Transportation: Barcelona has invested in intelligent transportation systems to improve mobility and reduce congestion. The city's bike-sharing program, supported by a mobile app, promotes sustainable transportation options, while smart parking systems help drivers locate available parking spaces efficiently.

Sustainable Building Design: Barcelona has made significant progress in adopting sustainable building practices. Several buildings in the city are equipped with smart technologies, including energy-efficient lighting, automated temperature control systems, and solar panels, reducing energy consumption and carbon footprint.

Citizen Engagement: Barcelona actively involves citizens in decision-making processes through digital platforms. The city's "Decidim" platform allows residents to contribute ideas, participate in public consultations, and collaborate with the government in shaping policies and initiatives.

Barcelona's smart city initiatives demonstrate the potential of technology to transform urban areas into sustainable and livable spaces. The city serves as a model for other urban centers seeking to leverage technology for sustainable development.

Conclusion

Technology plays a crucial role in shaping sustainable urbanization and transportation. It enables cities to achieve environmental, social, and economic sustainability by improving energy efficiency, enhancing mobility, and optimizing resource management. However, there are challenges that need to be addressed, such as affordability, the digital divide, and data privacy. By embracing technology and adopting an integrated approach, cities can create vibrant and resilient urban areas that enhance the quality of life for all residents while mitigating the impacts of climate change. The case study of Barcelona's smart city initiatives illustrates the transformative potential of technology in creating sustainable cities of the future.

Investigating the role of technology in promoting sustainable agriculture and food security

In recent years, the global population has been growing at an unprecedented rate, leading to increased demand for food. At the same time, the agricultural sector is facing numerous challenges such as climate change, water scarcity, and diminishing arable land. To address these challenges and ensure food security for future generations, technology plays a crucial role in promoting sustainable agriculture. In this section, we will explore the various ways in which technology can be utilized to improve agricultural practices, enhance productivity, and ensure food security.

Challenges in agriculture

Before delving into the role of technology, it is important to understand the challenges faced by the agricultural sector. These challenges include:

1. **Climate Change:** Increasing temperatures, erratic rainfall patterns, and extreme weather events pose a significant threat to crop yields and agricultural productivity. Technology can help develop innovative solutions to mitigate the impacts of climate change on agriculture.

2. **Water Scarcity:** Agriculture accounts for a significant portion of global water usage, and with increasing water scarcity, efficient water management is crucial. Technology can aid in developing water-saving irrigation systems and precision agriculture techniques.

3. **Land Degradation:** Soil erosion, deforestation, and urbanization are leading to the loss of arable land. Technology can play a vital role in soil conservation and land management practices to ensure sustainable agricultural production.

4. **Pest and Disease Management:** Pests, diseases, and weeds can cause significant crop losses. Technology can assist in early detection, monitoring, and control of pests and diseases, reducing the reliance on chemical pesticides.

5. **Unsustainable Farming Practices:** Traditional farming practices often lead to soil depletion, excessive use of fertilizers, and overuse of natural resources. Technology can provide alternatives such as precision farming, agroforestry, and organic farming, promoting sustainable agricultural practices.

Technological solutions

To address these challenges, several technological solutions can be implemented in sustainable agriculture and food security efforts. Let's explore some of the key technological innovations in this field:

1. **Precision Agriculture:** Precision agriculture utilizes advanced technologies such as remote sensing, geographic information systems (GIS), and global positioning systems (GPS) to collect data on soil properties, weather conditions, and crop growth. This data helps farmers make informed decisions regarding irrigation, fertilizer application, and pest control, reducing resource wastage and increasing crop yields.

2. **Genetic Engineering:** Genetic engineering plays a crucial role in developing crops with desirable traits such as drought resistance, disease resistance, and increased nutritional value. By utilizing biotechnology, researchers can enhance crop productivity and improve the nutritional content of staple crops.

3. **Smart Irrigation Systems:** Traditional irrigation methods often result in water wastage due to over-irrigation or inefficient water distribution. Smart irrigation systems use sensors and weather data to optimize water usage, delivering the right amount of water at the right time, ensuring efficient crop irrigation and water conservation.

4. **Vertical Farming:** With the increase in urbanization, arable land is becoming scarce. Vertical farming, a method that utilizes stacked layers to grow crops indoors, offers a solution to this problem. By utilizing controlled environments and LED lights, vertical farming can produce crops year-round with significantly reduced water usage and transportation costs.

5. **Drones and Robotics:** Drones equipped with remote sensing technology can capture high-resolution images of crop fields, providing valuable information on crop health and identifying areas of concern. On the other hand, agricultural robots can perform tasks such as planting, harvesting, and weeding autonomously, reducing the dependency on manual labor and increasing efficiency.

6. **Blockchain Technology:** Blockchain technology can enhance transparency and traceability in the food supply chain. By utilizing blockchain, consumers can have access to reliable information about the origin, handling, and quality of food products, ensuring food safety and reducing fraud.

Case Study: Digital Agriculture in India

India, as one of the world's largest agricultural economies, is adopting various technology-driven approaches to promote sustainable agriculture and food security. The implementation of digital agriculture practices has shown promising results in addressing the challenges faced by smallholder farmers. Let's take a closer look at some key initiatives:

1. **Soil Health Cards:** The Soil Health Card scheme aims to provide soil nutrient analysis-based recommendations to farmers, enabling them to make informed decisions about fertilizer application. This helps reduce excessive fertilizer usage and improve soil health.

2. **e-NAM:** The National Agricultural Market (e-NAM) is a pan-India electronic trading portal that connects agricultural markets, facilitating online trading of agricultural commodities. This initiative improves market access for farmers, reduces intermediaries, and ensures fair prices.

3. **Crop Insurance:** The Pradhan Mantri Fasal Bima Yojana (Crop Insurance Scheme) leverages remote sensing technology and satellite imagery to assess crop health and estimate crop losses due to natural disasters. This helps farmers obtain insurance coverage and financial support in times of need.

4. **Agri-Tech Startups:** The Indian agricultural technology startup ecosystem has witnessed significant growth in recent years. Startups are leveraging technology to address various challenges, including farm management, supply chain optimization, and market linkages.

Conclusion

Technology plays a pivotal role in promoting sustainable agriculture and ensuring food security in the face of numerous challenges. Precision agriculture, genetic engineering, smart irrigation systems, vertical farming, drones and robotics, and blockchain technology are among the key innovations driving sustainable agricultural practices. These technologies enable efficient resource management, enhance crop productivity, reduce environmental impacts, and improve farmers' livelihoods. As we move towards a more technology-driven future, it is essential to continue exploring and adopting innovative solutions to achieve sustainable agriculture and food security on a global scale.

Assessing the Potential of Digital Technologies for Monitoring and Managing Natural Resources

In today's era of increasing environmental concerns, the effective monitoring and management of natural resources is of paramount importance. Digital technologies have emerged as powerful tools that can revolutionize the way we monitor and manage our natural resources. In this section, we will explore the potential of digital technologies in this context, discussing their applications, benefits, and challenges.

Introduction to Digital Technologies for Natural Resource Management

Digital technologies encompass a diverse range of tools and systems that utilize computer technology, data analytics, and connectivity to collect, process, and analyze information. When applied to natural resource management, these technologies offer innovative solutions to address challenges such as resource depletion, environmental degradation, and biodiversity loss.

Applications of Digital Technologies in Natural Resource Monitoring

Digital technologies have a wide range of applications in monitoring natural resources, enabling real-time data collection, analysis, and decision-making. Some key applications include:

- **Remote Sensing**: Remote sensing technologies, such as satellite imaging and aerial drones, provide detailed and accurate data on land cover, vegetation health, and water quality. This information is crucial for monitoring deforestation, land degradation, and changes in biodiversity.

- **Internet of Things (IoT)**: IoT devices, including sensors and smart meters, can be deployed in various natural environments to collect data on water quality, air pollution, and soil moisture. This real-time data helps in identifying environmental risks and guiding effective resource management strategies.

- **Geographic Information Systems (GIS)**: GIS platforms integrate spatial data, satellite imagery, and other forms of geospatial data to map and analyze natural resources. These tools enable scientists and policymakers to visualize and understand complex ecological systems, aiding in decision-making and sustainable land use planning.

- **Data Analytics and Machine Learning:** Advanced data analytics and machine learning algorithms can process vast amounts of environmental data, identifying patterns, predicting changes, and optimizing resource management strategies. These technologies enable more efficient and targeted interventions for conservation and sustainable use of natural resources.

Benefits of Digital Technologies in Natural Resource Management

The adoption of digital technologies for natural resource monitoring and management offers several key benefits:

- **Improved Accuracy and Precision:** Digital technologies provide high-resolution data, eliminating errors associated with manual data collection. This accuracy enables more reliable and informed decision-making, leading to more effective resource management.

- **Real-time Monitoring and Early Warning Systems:** Digital technologies enable real-time monitoring of natural resources, allowing for early detection of environmental changes or risks. For example, IoT sensors can quickly detect water contamination, enabling timely interventions to prevent damage to ecosystems and human health.

- **Efficient Resource Allocation:** With the ability to collect and analyze large-scale data, digital technologies help optimize resource allocation. This leads to more efficient and sustainable use of natural resources, reducing waste and ensuring their long-term availability.

- **Enhanced Collaboration and Knowledge Sharing:** Digital platforms facilitate collaboration and knowledge sharing among scientists, policymakers, and communities. This interdisciplinary approach promotes the exchange of best practices and innovative solutions for sustainable natural resource management.

Challenges and Considerations

While the potential of digital technologies for monitoring and managing natural resources is immense, several challenges and considerations need to be addressed:

- **Data Privacy and Security:** The collection and storage of large-scale environmental data raise concerns about data privacy and security.

Measures must be in place to ensure the protection of sensitive information and prevent unauthorized access or misuse.

- **Data Integration and Interoperability:** Different digital technologies generate large volumes of data that need to be integrated and shared across platforms. Standardized protocols and interoperable systems are essential to avoid data fragmentation and enable effective data analysis.

- **Cost and Infrastructure:** The adoption of digital technologies requires significant investment in infrastructure development, equipment procurement, and skilled personnel. Particularly in developing countries, these costs can hinder widespread implementation and access to these technologies.

- **Capacity Building and Awareness:** To fully harness the potential of digital technologies, capacity building initiatives are needed to train stakeholders in data collection, analysis, and interpretation. Additionally, raising awareness about the benefits and opportunities of digital technologies is essential for their widespread adoption.

Case Study: Digital Technologies for Forest Monitoring

One notable application of digital technologies in natural resource management is forest monitoring. Satellite imaging, drone surveillance, and machine learning algorithms are being used to track deforestation, identify illegal logging activities, and assess forest carbon stocks. These tools provide decision-makers with accurate and up-to-date information, facilitating targeted interventions for forest conservation and sustainable management.

For example, in the Amazon rainforest, remote sensing technologies have been instrumental in monitoring deforestation rates and identifying areas at high risk of illegal activities. Real-time data from satellite imagery enables authorities to take immediate action to combat deforestation and enforce forest protection policies.

Conclusion

Digital technologies have the potential to revolutionize the monitoring and management of natural resources. By facilitating real-time data collection, analysis, and decision-making, these technologies enable more accurate, efficient, and sustainable resource management strategies. However, challenges such as data privacy, infrastructure development, and capacity building must be addressed for widespread adoption and equitable access to these technologies. With the right approach, digital technologies can play a critical role in ensuring the long-term conservation and sustainable use of natural resources.

Analyzing the challenges and opportunities of technology in tackling social inequalities

In today's rapidly evolving technological landscape, the phenomenon of social inequality has become a pressing concern. While technology has the potential to exacerbate existing inequalities, it also presents opportunities to address and tackle these disparities. This section will explore the challenges and opportunities that arise from the intersection of technology and social inequalities, with a focus on how technology can be leveraged to promote inclusivity and bridge the gap between different social groups.

Understanding Social Inequalities

Before delving into the challenges and opportunities, it is essential to have a clear understanding of social inequalities. Social inequalities refer to the unequal distribution of resources, opportunities, and privileges within a society. These inequalities can manifest in various forms, such as income disparities, educational gaps, healthcare access, and digital divides. They often stem from systemic factors, including socioeconomic status, gender, race, ethnicity, and geographical location.

Challenges and Risks of Technological Solutions

While technology holds promise in addressing social inequalities, it is crucial to acknowledge the challenges and risks that come with it. Here are some key challenges to consider:

1. **Accessibility and Affordability:** Technological advancements, such as smartphones, internet connectivity, and digital services, have become integral parts of daily life. However, access to these technologies remains unequal, particularly for marginalized communities. Limited access to affordable internet connectivity and devices can further deepen social inequalities.

2. **Technological Illiteracy:** Bridging the digital divide requires not only access to technology but also the necessary skills to utilize it effectively. Technological illiteracy, especially among older adults and disadvantaged groups, hinders their

ability to take advantage of digital platforms and services, perpetuating social inequalities.

3. **Bias and Discrimination:** Technology is not neutral and can reflect the biases ingrained in society. As algorithms and artificial intelligence play an increasingly significant role in decision-making processes, biases can perpetuate discriminatory practices and reinforce social inequalities.

4. **Data Privacy and Security:** The collection and utilization of personal data raise concerns about privacy and security. Inadequate protection measures can disproportionately affect marginalized individuals and communities who may already be vulnerable to exploitation.

5. **Automation and Job Displacement:** While automation can improve productivity and efficiency, it also has the potential to displace traditional jobs. If not managed carefully, this disruption can exacerbate social inequalities, leaving certain segments of the population at a greater disadvantage.

Opportunities and Strategies for Promoting Social Inclusion

Despite the challenges, technology offers various opportunities to address social inequalities and promote social inclusion. Here are some strategies and examples of how technology can be leveraged:

1. **Digital Inclusion Initiatives:** Governments, in collaboration with private sector partners, can implement initiatives to improve digital access and literacy among disadvantaged communities. This can involve subsidizing internet services, providing affordable devices, and offering training programs to bridge the digital divide.

2. **Education and Skills Development:** Technology can revolutionize education and offer opportunities for skill development, regardless of geographical location or socioeconomic status. Online learning platforms, open educational resources, and Massive Open Online Courses (MOOCs) enable individuals to access quality educational materials and acquire new skills.

3. **Social Entrepreneurship:** Social entrepreneurs can leverage technology to develop innovative solutions that address social inequalities. For instance, initiatives focusing on affordable healthcare, sustainable agriculture, and renewable energy can contribute to reducing disparities and improving the overall quality of life.

4. **Data-Driven Decision Making:** By harnessing the power of data analytics and artificial intelligence, policymakers can gain insights into societal challenges and design targeted interventions. Data-driven decision-making can lead to evidence-based policies that effectively address social inequalities.

5. **E-Government and Civic Engagement:** Technology can enhance citizen engagement and participation in decision-making processes. E-government platforms and digital tools enable individuals to voice their concerns, provide feedback, and hold governments accountable, thus promoting transparency and inclusivity.

6. **Social Media and Grassroots Movements:** Social media platforms play a crucial role in connecting like-minded individuals and organizing grassroots movements for social change. They can amplify marginalized voices, raise awareness about social inequalities, and mobilize collective action.

Case Studies and Best Practices

Examining successful case studies can provide valuable insights into how technology can effectively tackle social inequalities. Here are two examples:

1. **M-Pesa:** M-Pesa, a mobile money transfer service, has revolutionized financial inclusion in Kenya. By leveraging existing mobile phone infrastructure, M-Pesa enables individuals, particularly those in rural areas, to access financial services, make transactions, and build savings.

2. **Khan Academy:** Khan Academy, an online educational platform, offers free access to quality educational resources. It has become a valuable tool for bridging educational gaps and empowering learners from around the world, providing equal opportunities for education.

Ethical Considerations

As technology plays a more prominent role in addressing social inequalities, it is crucial to consider the ethical implications. Here are some key ethical considerations:

1. **Privacy and Consent:** Protecting individuals' privacy rights and obtaining informed consent for data collection and utilization are essential to avoid further marginalizing vulnerable populations.

2. **Accountability and Transparency:** Ensuring accountability and transparency in algorithmic decision-making processes is crucial to prevent biases and discrimination. Auditing algorithms and promoting algorithmic transparency can minimize the risks of perpetuating social inequalities.

3. **Inclusivity and Accessibility:** Designing technologies that are inclusive and accessible to all individuals, regardless of their abilities or circumstances, is essential to prevent the creation of new forms of inequalities.

4. **Equity and Fairness:** Striving for equity and fairness should be guiding principles when developing and implementing technological solutions. It is important to consider the differential impact of technology on different social groups and ensure that the benefits are distributed equitably.

Conclusion

While social inequalities pose significant challenges, technology presents numerous opportunities to address and tackle these disparities. By recognizing the challenges, implementing inclusive strategies, and considering ethical implications, technology can be a powerful tool in promoting social inclusion, bridging the divide, and creating a more equitable society.

Exercise: Research and discuss a recent case study where technology has been used to tackle a specific social inequality. Analyze the outcomes, challenges encountered, and potential for scalability and replication.

Further Reading:

1. Acemoglu, D., Autor, D., Dorn, D., Hanson, G., & Price, B. (2014). "Return of the Solow Paradox? It's the Ongoing Automation of Manufacturing". MIT Economics.

2. World Bank (2020). "Digital Economy for Africa: Challenges and Opportunities".

3. World Economic Forum (2020). "The Global Gender Gap Report 2020".

With the diligent application of technology and a focus on promoting inclusivity, we can work towards overcoming social inequalities and create a more equitable future for all members of society.

Understanding the role of technology in promoting resilient infrastructure and disaster management

Resilient infrastructure and effective disaster management are crucial components for ensuring the safety, well-being, and sustainable development of societies. Technology plays a significant role in enhancing the resilience of infrastructure and improving disaster management efforts. In this section, we will explore how various technological advancements contribute to these areas and enable more effective responses to disasters.

Importance of resilient infrastructure

Infrastructure refers to the basic physical and organizational structures needed for the functioning of a society or enterprise. Resilient infrastructure refers to infrastructure systems that are designed to withstand, adapt to, and quickly recover from disruptive events such as natural disasters, climate change impacts, and human-induced hazards. Resilient infrastructure is crucial for minimizing the negative impacts of disasters, ensuring the continuity of essential services, and facilitating rapid recovery.

The role of technology in promoting resilient infrastructure is multi-faceted. Technological innovations and advancements are instrumental in improving the design, construction, and operation of infrastructure systems to enhance their resilience. Let's explore some key areas where technology contributes to resilient infrastructure and disaster management.

Monitoring and early warning systems

One of the fundamental aspects of disaster management is monitoring and early warning systems. Technology plays a vital role in the development and implementation of these systems. Various sensors, remote sensing technologies, and data analytics techniques are used to monitor environmental conditions, detect potential hazards, and issue timely warnings to vulnerable communities.

For example, in the case of natural disasters like hurricanes or earthquakes, early warning systems utilize seismic sensors, weather satellites, and data analysis algorithms to detect and predict the events. Advanced modeling techniques combined with real-time data can provide accurate information about the magnitude, intensity, and potential impacts of the disaster. This information can help authorities and communities make informed decisions, evacuate vulnerable areas, and allocate resources more efficiently.

Information and communication technology (ICT) infrastructure

Effective communication during a disaster is vital for coordinating response efforts, disseminating information to affected populations, and facilitating the rescue and recovery operations. Robust information and communication technology (ICT) infrastructure is an essential enabler in disaster management.

In the aftermath of a disaster, traditional communication channels such as telephone lines may be disrupted or overwhelmed. However, modern technologies like mobile networks, satellite communications, and internet-enabled systems provide alternative means of communication that can remain operational even in challenging circumstances. These technologies ensure that emergency services, relief organizations, and affected individuals can stay connected and coordinate effectively.

Moreover, ICT infrastructure enables the collection and analysis of vital data during disaster events. This data can include real-time weather updates, evacuation routes, medical facilities' availability, and damage assessments. By integrating this data into decision support systems, authorities can make informed decisions and allocate resources more effectively.

Remote sensing and geospatial technologies

Remote sensing and geospatial technologies are essential tools in disaster management and infrastructure resilience. These technologies involve the collection, processing, and analysis of information about the Earth's surface and its features from a distance, often using satellites or aircraft.

Remote sensing enables the monitoring of changes in the landscape, such as deforestation, urban growth, or land subsidence, which can have significant implications for infrastructure resilience. By analyzing remote sensing data, authorities can identify areas at risk, evaluate the vulnerability of critical infrastructure, and take appropriate measures to enhance resilience.

Geospatial technologies, on the other hand, allow for the mapping and spatial analysis of various data layers, such as population density, critical infrastructure locations, and hazard zones. By overlaying these layers, decision-makers can assess the potential impacts of disasters on infrastructure and plan for resilient development. Furthermore, geospatial technologies aid in monitoring and managing post-disaster recovery efforts and reconstruction.

Building materials and construction techniques

Innovations in building materials and construction techniques are essential for enhancing the resilience of infrastructure. Technology has contributed to the development of stronger, more durable, and disaster-resistant materials and construction practices.

For instance, in areas prone to earthquakes, engineers have developed seismic-resistant building techniques that can withstand the shaking forces and avoid structural failure. These include the use of reinforced concrete, base isolation systems, and energy dissipation devices. Similarly, for regions prone to hurricanes or high wind speeds, building designs incorporate advanced wind engineering principles and the use of impact-resistant materials.

Moreover, technological advancements have led to the development of smart infrastructure systems that can self-monitor, detect damage, and respond accordingly. For example, sensors embedded in buildings and bridges can detect structural weaknesses or damage and send real-time alerts to maintenance teams, enabling timely repairs and preventing further deterioration.

Resilient energy systems

Reliable and resilient energy systems are critical for supporting disaster management efforts and ensuring the functioning of infrastructure during and after a disaster. Technology plays a significant role in enhancing the resilience of energy systems.

Renewable energy technologies, such as solar and wind power, contribute to the resilience of energy systems by diversifying the energy sources and reducing dependence on centralized grids. Distributed generation systems, combined with energy storage technologies, enable communities to maintain power supply even if the main grid is disrupted. Microgrids, which are small-scale localized energy systems, provide an additional layer of resilience by allowing for localized power generation and distribution.

Furthermore, digital technologies and smart grid systems enable real-time monitoring and control of energy supply and demand. These systems facilitate quick identification of faults, rerouting of power, and effective load management, improving the resilience and reliability of energy infrastructure.

Challenges and opportunities

While technology presents numerous opportunities for promoting resilient infrastructure and disaster management, several challenges need to be addressed to ensure effective implementation. These challenges include:

- Cost-effectiveness: Many advanced technologies and solutions may come with high costs, limiting their adoption in low-resource settings and developing countries. Balancing the cost-effectiveness of technologies and their potential benefits is crucial.

- Technical capacity: Building the technical capacity to develop, operate, and maintain advanced technology systems requires specialized skills and knowledge. Ensuring adequate training, education, and knowledge transfer is essential.

- Interoperability and data sharing: Effective disaster management relies on the integration and sharing of data across various systems and stakeholders. Ensuring interoperability and addressing concerns related to privacy and data security are significant challenges.

- Accessibility and inclusiveness: Technology should be accessible to all segments of society. Ensuring inclusiveness and considering the needs of marginalized communities, people with disabilities, and those with limited access to technology is crucial.

Addressing these challenges requires collaboration among various stakeholders, including governments, industry, academia, and communities. By promoting research and development, fostering innovation, and investing in technological infrastructure, societies can enhance their resilience, reduce the impacts of disasters, and build a sustainable future.

Key Takeaways

- Technology plays a significant role in promoting resilient infrastructure and effective disaster management. - Monitoring and early warning systems, ICT infrastructure, remote sensing and geospatial technologies, building materials and construction techniques, and resilient energy systems are some areas where technology contributes. - Challenges include cost-effectiveness, technical capacity, interoperability and data sharing, and accessibility and inclusiveness. - Collaboration among stakeholders is crucial for addressing these challenges and harnessing the potential of technology to build resilience.

Examining the implications of technological advancements on biodiversity and ecosystem preservation

Technological advancements have significantly impacted various aspects of society, including the natural environment. In recent years, there has been growing concern about the implications of these advancements on biodiversity and ecosystem preservation. This section will explore the potential consequences of technology on ecosystems, as well as the ways in which technology can be harnessed to protect and conserve biodiversity.

The relationship between technology and biodiversity

Biodiversity refers to the variety and abundance of living organisms in a given area. It encompasses the diversity of species, genetic diversity within species, and the diversity of ecosystems. Technology has both positive and negative impacts on biodiversity.

One positive impact of technology on biodiversity is the ability to monitor and study ecosystems more comprehensively. Technological advancements in remote sensing, satellite imagery, and geographic information systems (GIS) have revolutionized the field of biodiversity monitoring and conservation. These tools allow scientists to assess the distribution and abundance of species, identify critical habitats, and monitor ecosystem changes over time.

However, technology can also pose significant threats to biodiversity. For example, habitat destruction and fragmentation caused by infrastructure development, such as roads and buildings, can lead to the loss of important habitats and disrupt ecological processes. Pollution resulting from the use of technology, such as industrial waste and chemical pollutants, also poses a significant threat to biodiversity.

Challenges and risks of technological advancements

While technology can provide valuable tools for biodiversity conservation, it is essential to recognize and address the challenges and risks associated with its use.

One challenge is the rapid pace of technological advancements. The speed at which new technologies are developed and implemented often outpaces our understanding of their potential impacts on biodiversity. This can result in unintended consequences, such as the spread of invasive species through global transportation networks or the accidental introduction of genetically modified organisms into ecosystems.

Another risk is the potential for technology to create dependencies and disrupt natural processes. For example, the use of artificial habitats or artificial pollination methods could lead to a decline in native species that rely on natural ecological interactions. Furthermore, the reliance on sophisticated technology for conservation efforts may exclude or disadvantage communities with limited access to resources and knowledge.

Harnessing technology for biodiversity conservation

Despite these challenges, technology also provides significant opportunities for biodiversity conservation and ecosystem preservation. Here are some examples of how technology can be harnessed for these purposes:

1. DNA barcoding: Advances in DNA sequencing technology have enabled the development of DNA barcoding, a technique that uses short DNA sequences to identify species. DNA barcoding can accelerate species identification, monitor illegal wildlife trade, and assist in the detection of invasive species.

2. Remote sensing and GIS: Remote sensing tools, such as satellite imagery, can provide valuable information on land cover, vegetation health, and habitat suitability. Coupled with GIS, these technologies enable the identification of critical habitats, monitoring of land-use changes, and evaluation of conservation strategies.

3. Conservation drones: Unmanned aerial vehicles (UAVs), or drones, can be equipped with cameras and sensors to survey and monitor ecosystems. Drones can reach remote or inaccessible areas, providing valuable data on wildlife populations, habitat conditions, and illegal activities, such as poaching or deforestation.

4. Citizen science: Technology can facilitate citizen engagement in biodiversity conservation through citizen science initiatives. Mobile applications and online platforms allow individuals to contribute to species monitoring, data collection, and environmental reporting, increasing public awareness and involvement in conservation efforts.

5. Conservation genetics: Advances in genetic technologies, such as genomic sequencing, can help inform conservation strategies. Understanding the genetic diversity and population structure of endangered species can guide breeding

programs, habitat restoration efforts, and the identification of genetically distinct populations for protection.

Ethical considerations

As with any use of technology, ethical considerations play a crucial role in harnessing technology for biodiversity conservation. It is important to ensure that the benefits of technological advancements are balanced with potential risks and unintended consequences.

One ethical consideration is the involvement of local communities in decision-making processes related to technology use in biodiversity conservation. Engaging local knowledge and perspectives can help ensure that technology is appropriately adapted to local contexts and that conservation efforts are inclusive and sustainable.

Additionally, issues of privacy and data security need to be addressed when collecting and analyzing data using technological tools. It is essential to protect sensitive information, such as the location of endangered species, to prevent potential harm to wildlife or exploitation by unauthorized individuals.

Case study: Using technology to combat wildlife trafficking

One notable example of technology being used for biodiversity conservation is the fight against wildlife trafficking. Illicit wildlife trade poses a significant threat to biodiversity, and technology has become a valuable tool in combating this illegal activity.

DNA barcoding techniques have been utilized to identify wildlife products and trace their origins accurately. By comparing DNA samples from confiscated products to genetic databases, law enforcement agencies can identify the species and geographic region from which the products originated. This information helps target enforcement efforts, disrupt supply chains, and hold perpetrators accountable.

Furthermore, the use of surveillance technology, such as camera traps, drones, and acoustic sensors, has improved monitoring and detection capabilities. These tools enable authorities to identify wildlife poaching hotspots, track illegal activities, and gather evidence for prosecution.

Conclusion

Technological advancements have both positive and negative implications for biodiversity and ecosystem preservation. While technology can pose risks, such as habitat destruction and pollution, it also provides valuable tools for monitoring, research, and conservation efforts. By harnessing technology ethically and strategically, we can leverage its potential to protect and conserve biodiversity for future generations.

Discussing Strategies for Fostering Responsible and Sustainable Technology-Driven Development

In this section, we will explore various strategies that can be adopted to promote responsible and sustainable technology-driven development. As technology continues to advance at a rapid pace, it is crucial to ensure that its development and implementation align with the principles of sustainability and contribute to the overall well-being of society. By adopting appropriate strategies, we can harness the potential of technology to address pressing global challenges while minimizing any adverse impacts.

1. Emphasizing the Triple Bottom Line

One effective strategy for fostering responsible and sustainable technology-driven development is to emphasize the concept of the triple bottom line. This approach recognizes that economic prosperity, environmental sustainability, and social well-being are interconnected and should be given equal consideration in decision-making processes.

Companies and organizations should evaluate the potential impacts of technology on each of these pillars. For example, when introducing a new technology, they should assess its environmental footprint, such as its energy consumption, resource usage, and waste generation. They should also consider the social implications, such as its potential to create or displace jobs, its accessibility to marginalized communities, and its potential to bridge the digital divide. Finally, they should evaluate the economic viability, considering factors such as cost-effectiveness, return on investment, and potential for long-term growth.

By adopting a triple bottom line approach, technology-driven projects can be designed and implemented in a way that maximizes benefits across these three dimensions, promoting responsible and sustainable development.

2. Promoting Collaboration and Partnerships

Another key strategy for fostering responsible and sustainable technology-driven development is to promote collaboration and partnerships among various stakeholders. This includes collaboration between governments, academia, industry, non-governmental organizations (NGOs), and local communities.

Through collaboration and partnerships, diverse perspectives can be brought together, fostering innovation and ensuring that technology-driven development aligns with the needs and aspirations of all stakeholders. For instance, academia can contribute by conducting research on the environmental and social impacts of different technologies, informing policy decisions. NGOs can play a critical role in advocating for responsible and sustainable technology adoption, ensuring that the concerns of local communities and vulnerable groups are taken into account.

Additionally, collaboration between governments and industry can facilitate the development of supportive policies and regulations that promote responsible technology adoption and incentivize sustainable practices. By working together, stakeholders can share knowledge, resources, and best practices, ultimately fostering responsible and sustainable technological development.

3. Integrating Circular Economy Principles

Integrating circular economy principles into technology-driven development is another crucial strategy for promoting responsible and sustainable practices. The circular economy aims to minimize waste and maximize resource efficiency by emphasizing product reuse, recycling, and regeneration.

In the context of technology, this could involve designing products with a longer lifespan, ensuring that they are easily repairable and upgradable, and facilitating the recycling and responsible disposal of electronic waste. By adopting circular economy principles, we can reduce the environmental impact of technology and promote a more sustainable and resilient economy.

For example, in the context of renewable energy technologies, adopting circular economy principles could involve designing solar panels with materials that can be easily recovered and reused at the end of their life cycle. This minimizes the environmental impact associated with the extraction and production of raw materials required for their manufacture.

4. Ensuring Access and Inclusion

Ensuring access and inclusion is a critical strategy for fostering responsible and sustainable technology-driven development. It is essential to bridge the digital divide and ensure that the benefits of technology are accessible to all, regardless of their socioeconomic status or geographical location.

Governments, in collaboration with other stakeholders, should work towards providing affordable and reliable internet connectivity to unserved and underserved areas. Efforts should also be made to enhance digital literacy and skills development, enabling individuals to fully participate in the digital economy. Inclusive technology design should be emphasized, considering the diverse needs and capabilities of users, including those with disabilities.

Moreover, promoting gender equality in technology-related fields is crucial for fostering responsible and sustainable development. Efforts should be made to bridge the gender gap in STEM education and careers, promoting diversity and ensuring that the perspectives and talents of women are fully integrated into the technological advancements of society.

5. Encouraging Ethical Considerations

Ethical considerations should be a fundamental part of technology-driven development. As technologies become more advanced and ubiquitous, it is crucial to examine their ethical implications and address potential risks and unintended consequences.

Organizations and governments should prioritize the development and implementation of ethical guidelines and frameworks to ensure responsible and accountable technology adoption. This could include principles such as privacy protection, transparency, algorithmic fairness, and accountability for the ethical use of emerging technologies such as artificial intelligence and blockchain.

Furthermore, promoting digital ethics education and awareness among individuals, organizations, and policymakers is vital. This will enable society to navigate the complex ethical dilemmas posed by technology and make informed decisions that prioritize the well-being of individuals and communities.

6. Investing in Research and Development for Sustainable Technologies

Investing in research and development (R&D) for sustainable technologies is a significant strategy for fostering responsible and sustainable technology-driven development. R&D efforts should focus on developing technologies that have a minimal environmental impact, promote social well-being, and contribute to economic growth.

Governments and organizations should allocate funding and resources to support R&D initiatives aimed at developing sustainable technologies. This includes research on renewable energy sources, energy-efficient technologies, waste management solutions, and low-carbon transportation systems, among others.

Additionally, collaboration between academia, industry, and government institutions in research and development can accelerate technological innovation and ensure that new technologies are aligned with sustainability goals.

Conclusion

Fostering responsible and sustainable technology-driven development requires a holistic approach that considers economic, environmental, and social dimensions. By emphasizing the triple bottom line, promoting collaboration, integrating circular economy principles, ensuring access and inclusion, encouraging ethical considerations, and investing in research and development, we can harness the power of technology to address global challenges in a sustainable and responsible manner. By adopting these strategies, we can shape a future where technology serves as a catalyst for positive change, contributing to a more equitable, resilient, and sustainable world.

Index

-effectiveness, 39, 107, 332, 335

ability, 21, 23, 42, 51, 107, 108, 114, 137, 148, 149, 151, 164, 177, 181, 219, 253, 260, 262, 272, 284, 286, 288, 290, 303, 308
absence, 139
absorption, 105
abundance, 332
abuse, 218, 239
academia, 13, 28, 60, 93, 94, 132, 155, 185, 208, 220, 228, 236, 240, 269, 293, 331, 336, 338
academic, 115, 205, 228, 262
acceleration, 296
accelerator, 189
access, 4, 6, 9, 20–25, 27, 29–32, 40, 43–46, 61, 66, 67, 70, 75, 76, 78, 92–95, 97, 103, 122, 125, 129–133, 135, 137, 138, 140, 142, 159, 160, 163, 164, 169, 170, 173, 180, 181, 183–187, 189, 197, 201, 202, 204–206, 208, 214–216, 218, 226, 231, 234, 235, 239, 257, 260, 261, 263, 265–269, 273–275, 277, 279, 283, 286, 287, 289–293, 295–298, 304, 324, 333, 336, 338
accessibility, 65, 102, 103, 137, 150, 171, 239, 290, 332, 335
account, 219, 236, 267, 279, 303, 311, 336
accountability, 27, 33, 34, 53, 131, 133, 141, 170, 236, 337
accounting, 187
accumulation, 305
accuracy, 53, 113
acquisition, 22, 180, 204, 206
action, 193, 323
activity, 106, 183, 228, 334
acumen, 181
adaptability, 17, 58, 147, 199
adaptation, 43, 57, 58, 169, 240, 246, 248, 270
addition, 26, 39, 111, 181, 226, 273
address, 4, 5, 8, 19, 22, 24–28, 32–34, 44, 48, 58, 59, 61, 67, 68, 76, 81, 91, 95, 119, 131–133, 137, 139, 141, 142, 148, 159, 169, 171, 177, 184, 185, 188, 200,

203, 207, 209, 210,
217–219, 226, 229,
235–237, 239, 240, 245,
248, 250–252, 254, 255,
257, 261, 263, 264, 266,
268, 269, 286, 287, 291,
294–296, 298, 299, 304,
306, 318, 319, 321, 324,
325, 327, 332, 335, 337,
338
adoption, 5–9, 15, 20, 24–32, 40,
46–50, 52, 53, 56, 58, 68,
79–81, 89, 92, 94, 98, 99,
103, 107, 110, 111, 115,
120, 130–132, 135–155,
160, 161, 164–167,
169–173, 180, 203, 220,
229, 236, 247, 279, 284,
290, 293, 294, 297, 299,
300, 304–308, 310, 312,
322, 324, 336, 337
advance, 3, 14, 16, 18, 112, 125, 167,
176, 203, 243, 254, 263,
283, 295, 335
advancement, 53, 93, 94, 98, 200,
260, 265, 274, 277, 279,
290
advantage, 79, 193, 203, 215, 275,
290
advent, 18, 41–44, 214, 235, 279,
284
advertising, 289, 295
advice, 25, 118
advocacy, 205
aerospace, 106, 247
affordability, 102, 136, 267, 290,
291, 317
Africa, 201
aftermath, 329

age, 20, 83, 122, 126, 148, 173, 214,
259–261, 265, 268, 290,
294
agent, 116
agriculture, 10, 36, 50, 78, 101, 103,
228, 229, 305, 307, 308,
318–320
agroforestry, 307
aid, 330
aim, 172, 217, 235, 236, 239, 263,
268, 298, 305, 309, 311
air, 40, 47, 76, 106, 108, 304, 306,
308
aircraft, 329
Alan Turing, 41
algorithm, 199, 247
Alibaba, 45, 199, 287, 288
alignment, 227
alleviation, 304
allocation, 2, 52, 68
alternative, 54, 140, 164, 245, 304,
329
aluminum, 79
amount, 2, 114, 249, 308
analysis, 26, 27, 43, 52, 131, 140,
235, 247, 253, 260, 321,
323, 329, 330
Andrew McAfee, 17
angel, 204
animal, 102
anonymity, 170
app, 248, 316
application, 10, 104–106, 118, 119,
148, 207, 227, 228, 305,
323, 328
approach, 5, 27, 46, 67, 78, 81, 101,
142, 167, 198, 199, 202,
209, 217, 224, 226, 236,
239, 252, 256, 261, 265,

Index

289, 300, 303, 317, 324, 335, 338
approval, 59
aquaculture, 307
area, 105, 106, 200, 218, 332
arena, 21
aspect, 92, 93, 238, 239, 268, 274, 275
assembly, 14, 39, 113, 245, 253, 276, 282
assessment, 26, 32, 139, 167, 236, 240
asset, 298
assistance, 187, 223, 292
association, 247
asteroid, 128
asymmetry, 22, 273
atmosphere, 54, 305, 306
attention, 103, 118, 169, 303
attraction, 305
attractiveness, 291
audience, 22, 200, 266, 284, 285, 298
auditing, 170
audits, 27, 296
Australia, 97
authentication, 286
authenticity, 50, 273
authorship, 214
automate, 3, 14, 42, 49, 70, 253
automation, 2, 3, 9, 15–18, 20, 21, 23, 27, 33, 39, 42, 45, 48, 50, 52–54, 112, 113, 118, 131, 204, 243–246, 249, 250, 252–254, 264, 272, 284
automobile, 245
autonomy, 34, 170

availability, 31, 38–40, 45, 54, 65, 68, 70, 78, 80, 122, 136, 137, 183, 185, 228, 257, 275, 290, 291, 329
avenue, 279
aversion, 183
awareness, 29–31, 40, 135, 142, 166, 181, 236, 333, 337

backbone, 75
background, 33, 41
balance, 58, 103, 170, 193, 194, 209, 213–216, 219, 239, 246, 256, 289, 295, 303
bandwidth, 286
Bangladesh, 25
bank, 181, 279
banking, 4, 20, 23, 25, 159, 160, 164, 279
Barcelona, 316, 317
barcoding, 333, 334
bargaining, 261
barrier, 24, 28–31, 139–141, 290, 297, 307
base, 24, 70, 199, 217, 275, 287, 330
basis, 248
behavior, 142, 239, 289
being, 24, 47, 55, 101, 116, 129, 135, 171, 210, 235, 240, 246, 291, 297, 303, 323, 328, 334, 335, 337
belief, 224
benefit, 15, 20, 21, 29, 33, 48, 79, 80, 118, 133, 140, 156, 230, 237, 250, 264, 268–270, 294, 298, 299
betterment, 34, 234
bias, 27, 33, 34, 49, 53, 131, 169, 170, 203, 236, 266, 269,

270
bike, 316
biodiversity, 103, 307, 321, 332–335
bioenergy, 112
biomedicine, 93
biotechnology, 60, 101–103, 130, 209, 294
blockchain, 50, 51, 118–122, 131, 204, 206, 209, 247, 273, 320, 337
board, 182
body, 105, 238
book, 38
border, 21, 50, 219, 271, 273, 284–289, 298
bottom, 335, 338
box, 177
brand, 181, 199, 214, 238
branding, 199
breach, 32, 125
break, 265
breakthrough, 228
breeding, 333
brick, 159
bridge, 5, 61, 66, 131, 137, 140, 170, 172, 220, 235, 252, 259, 261, 263, 268, 290, 296, 298, 324, 335–337
broadband, 92–94, 185, 268, 286
broadcasting, 127
budget, 26
building, 40, 81, 92–94, 138, 181, 199, 214, 234, 285, 286, 298, 307, 316, 323, 330, 332
burden, 28, 137, 142
burning, 305
business, 2, 20, 41, 42, 44, 52, 67, 83, 138, 173, 177, 180, 181, 184–187, 192, 193, 199, 204, 205, 215, 219, 246, 261, 272, 273, 276, 279, 284
buyer, 287

California, 180
camera, 334
campaign, 199
cancer, 101
capability, 27, 147
capacity, 65, 76, 79, 106, 234, 286, 298, 307, 323, 332
capita, 2
capital, 24, 30, 93, 95, 97, 180–183, 186, 194, 195, 204, 206
capture, 199, 306
car, 245
carbon, 34, 40, 47, 51, 54, 55, 77, 80, 81, 104, 106, 111, 112, 130, 236, 296, 305, 308, 311, 312, 316, 323
care, 28, 47, 118, 149, 150
career, 264, 265
cargo, 76
case, 31, 53, 60, 71, 77, 82, 88, 92–94, 100, 110, 142, 160, 164, 166, 169, 172, 173, 175, 176, 192, 198, 200, 201, 208, 227, 230, 256, 257, 263, 269, 270, 276, 287, 317, 326, 329
cash, 159, 181
cashier, 15
catalyst, 2, 6, 9, 14, 249, 265, 272, 338
cement, 79, 81
century, 35, 41
certification, 34, 300

chain, 4, 22, 42, 50, 53, 70, 192, 204, 247, 272, 275–277, 282, 296, 300
challenge, 22, 26, 28, 33, 76, 77, 107, 131, 177, 180, 204, 205, 214, 216, 218, 219, 255, 273, 279, 290, 291, 307, 333
change, 10, 17, 22, 23, 26–28, 30, 40, 46, 47, 51, 54, 67, 132, 139, 141, 142, 169, 177, 200, 203, 207, 229, 236, 260, 262, 265, 267, 270, 299, 301, 303–305, 312, 317, 318, 328, 338
characteristic, 288
characterization, 104
charge, 105
checkout, 15
chemical, 79, 332
child, 300
China, 77, 81, 82, 199, 274, 276, 282
China, 276
choice, 239
citizen, 333
city, 316, 317
clarity, 141
clean, 4, 25, 47, 49, 54, 55, 67, 80, 81, 107, 108, 110, 111, 201, 209, 229, 294, 300, 303, 305–308
clearance, 21, 273
climate, 40, 47, 51, 54, 67, 77, 207, 229, 236, 299, 303–305, 312, 317, 318, 328
cloud, 45, 139, 166, 199, 284, 285
co, 285
coal, 36

cohesion, 55, 171
collaboration, 13, 14, 28, 42, 58, 60, 65, 88, 93, 129, 131–133, 141, 155, 156, 160, 161, 166, 173, 180, 184–186, 193, 196, 197, 205, 206, 208, 209, 219, 220, 223, 224, 226–228, 230, 234, 239, 240, 246, 247, 252, 253, 256, 261–264, 266, 270, 272, 274, 276, 285, 288, 289, 293, 298, 300, 312, 331, 336–338
collateral, 181
collection, 27, 32, 131, 235, 295, 298, 321, 323, 329, 333
collusion, 219, 239
color, 169
combination, 13, 19, 56, 200
comfort, 311
commerce, 4, 24, 44, 45, 48, 70, 192, 199, 215, 248, 266, 284, 287, 288
commercialization, 107, 151, 181, 208, 227–231
commitment, 31, 93, 193
communication, 4, 9, 27, 32, 38–40, 42, 44, 71, 75, 98, 127, 131, 135, 136, 139, 141, 233, 253, 260, 262, 263, 272, 274–276, 284, 285, 288, 304, 329
community, 55, 172, 173, 265
company, 25, 125, 164, 182, 183, 192, 193, 198, 199, 226, 282
compatibility, 26, 28, 138, 293
competence, 148
competition, 46, 59, 180, 194, 199,

205, 214, 216–220, 236,
    239, 274, 289, 299, 300
competitiveness, 21, 22, 66, 76, 78,
    79, 81, 88, 92, 113, 128,
    142, 207, 227, 243, 272,
    289
competitor, 43
complex, 6, 15, 18, 23, 34, 41, 46,
    47, 49, 52, 103, 107, 113,
    114, 116, 133, 135, 142,
    160, 205, 213, 215, 218,
    219, 226, 230, 246, 250,
    253, 261, 263, 275, 276,
    282, 284, 303, 304, 337
complexity, 226, 227
compliance, 118, 140, 141, 205, 226,
    273, 285, 286
component, 40, 67, 78, 85, 183
composition, 3, 15, 252–254
computer, 18, 41, 42, 51, 115, 116,
    122, 172, 257, 262, 321
computing, 34, 45, 166, 199, 284,
    285
concentration, 131, 180, 219, 239,
    260, 289, 299
concept, 44, 98, 257, 268, 303, 335
concern, 32, 48, 54, 122, 218, 263,
    294, 324, 332
conclusion, 9, 14, 34, 40, 48, 51, 67,
    75, 138, 141, 145, 177,
    183, 186, 193, 206, 210,
    234, 261, 274, 276, 289,
    300, 308
concrete, 330
conductivity, 106
conferencing, 42, 272
confidence, 125, 126, 147, 148, 181,
    208
confidentiality, 122, 215, 226

confusion, 214
congestion, 76, 77, 316
conglomerate, 199
connection, 18
connectivity, 4, 20, 22, 24, 31, 40,
    44–46, 50, 66, 67, 69, 71,
    73, 76, 77, 82, 89, 91, 93,
    94, 97, 126, 127, 130, 132,
    137, 183, 185, 265, 267,
    268, 286, 296, 321, 337
consent, 32, 50, 169, 298
consequence, 246
conservation, 80, 307, 316, 323, 324,
    332–335
consideration, 51, 96, 97, 115, 116,
    169, 217, 304, 334, 335
consolidation, 218
construction, 67, 77, 80, 106, 328,
    330, 332
consultation, 170
consulting, 284
consumer, 21, 46, 58, 194, 214, 215,
    238, 239, 272, 274, 289
consumption, 34, 40, 104, 204, 296,
    308, 311, 316, 335
content, 102, 173, 214, 247, 290,
    291, 297, 298
context, 34, 38, 66, 95, 190, 217,
    264, 299, 321, 336
continuity, 276, 328
contract, 247
contrast, 200
contribute, 4, 6, 9, 18, 19, 23, 25, 32,
    34, 70, 77, 94, 107, 112,
    139, 152, 164, 172, 173,
    189, 190, 193, 200, 205,
    214, 227, 236, 260, 261,
    264, 269, 270, 295,
    298–300, 304, 314, 316,

Index 345

    328, 330, 333, 335, 336
contribution, 178
contributor, 20, 306
control, 3, 33, 101, 102, 104, 115,
    116, 182, 213, 214, 234,
    235, 253, 305–307, 316,
    331
convenience, 44, 45, 135, 159
conversion, 98, 105, 107
cooking, 40
cooperation, 22, 66, 215, 219, 240,
    292, 293, 295, 298
coordination, 4, 22, 28, 223, 247,
    275
copying, 214
copyright, 214, 238
core, 98, 303
cornerstone, 2
corruption, 31, 131
cost, 4, 21, 24, 28, 39, 56, 79, 95, 97,
    107, 112, 116, 130, 136,
    140, 142, 193, 226, 243,
    245, 259, 267, 272, 275,
    276, 282–285, 287, 332,
    335
counsel, 205
counseling, 256
counterfeiting, 50, 215
country, 4, 54–56, 66, 69, 70,
    75–77, 81, 92–94, 97,
    141, 156, 160, 185, 230,
    257, 272, 276, 279
cover, 218, 305, 333
coverage, 127, 136
creation, 2–4, 6, 10, 14–18, 23,
    46–49, 51, 52, 67, 76, 83,
    119, 160, 180, 198, 207,
    214, 218, 227, 228, 231,
    246–248, 257, 260, 272,
    273, 285, 291, 295, 305,
    308, 314
creativity, 3, 13, 15, 52, 147, 216,
    263, 285
creator, 214
credibility, 181, 183
credit, 125, 159, 164
crop, 101, 102, 229, 307, 320
cross, 50, 131, 219, 228, 231, 266,
    269, 271, 273, 284–289,
    298
crowdfunding, 204, 266
culture, 13, 14, 16, 17, 27, 28, 58,
    94, 139, 150, 180, 184,
    205, 214, 227, 228, 254
currency, 272
curriculum, 94, 150, 263
customer, 21, 24, 42, 44, 48, 49, 70,
    114, 185, 199, 205, 215,
    227, 243, 246–248, 273,
    275, 276, 285, 287
customization, 114, 116
cutting, 95, 101
cyber, 27, 48, 122, 123, 125
cybersecurity, 14, 22, 23, 27, 44, 46,
    48–50, 94, 116, 122–125,
    131, 132, 141, 204, 247,
    264, 273, 286, 287, 295
cycle, 227, 336
cycling, 106

damage, 122, 329, 330
data, 15, 21, 22, 27, 28, 32–34, 40,
    42, 44, 45, 48, 50, 52, 71,
    82, 98, 105, 112, 114–116,
    118, 122, 125, 128, 129,
    131, 132, 140, 141, 149,
    163, 169, 170, 199, 204,
    205, 218, 219, 235, 247,

248, 253, 256, 260, 264,
272, 273, 284–290, 295,
298, 300, 317, 321, 323,
329, 330, 332–334
date, 31, 213, 323
day, 187, 189
dealing, 218
debate, 16, 103
decision, 2, 21, 27, 33, 42, 43, 45,
48, 51, 52, 114, 133, 139,
142, 167, 170, 199, 204,
219, 235, 236, 289, 303,
304, 316, 321, 323, 329,
330, 334, 335
decline, 3, 253, 333
decrease, 15, 47, 243
defect, 115
defense, 122
deforestation, 306, 323, 329
degradation, 77, 105, 296, 304, 321
delivery, 48, 53, 63, 93, 95, 105, 107,
129, 149, 166, 173, 204,
248, 276, 284–287, 295
demand, 3, 15, 17, 45, 48, 52, 182,
243, 246, 247, 252–254,
260, 276, 311, 318, 331
demo, 187, 189
density, 330
dependence, 39, 54, 79, 122, 275,
330
dependency, 67, 305
depletion, 40, 236, 304, 321
deployment, 32, 49–51, 53, 54, 80,
97, 108, 112, 118, 127,
133, 170, 294, 304, 305,
307, 308
deposition, 114
design, 16, 34, 77, 97, 105, 114, 128,
135, 173, 181, 198, 247,
267, 282, 299, 304, 313,
328, 337
designing, 105, 116, 123, 138, 263,
279, 336
destruction, 332, 335
detect, 47, 115, 329, 330
detection, 50, 106, 118, 170, 193,
333, 334
deterioration, 330
determinant, 68
development, 1–3, 5, 6, 9, 13, 16, 17,
20, 22–24, 27, 28, 30, 32,
36, 38–41, 44–47, 51–56,
58, 60, 63, 65–68, 70–72,
75–78, 80–83, 85, 88–95,
97, 98, 100–111, 114, 116,
122, 123, 126–133, 135,
137–139, 141, 142, 146,
147, 150, 151, 155–160,
163, 166, 170, 172, 173,
177, 180–186, 190,
193–195, 197–199, 204,
205, 207–210, 214, 216,
223, 226–228, 231, 236,
239, 240, 247, 248, 254,
261–264, 268, 271, 274,
276, 277, 279, 280, 286,
288–292, 294, 295,
297–301, 303–305, 307,
308, 313, 317, 323, 328,
330–333, 335–338
device, 228
diabetes, 101
diagnose, 49, 53
diagnosis, 47, 105, 118, 129
diagnostic, 101, 103, 130
dialogue, 293
difference, 42
diffusion, 10, 28–32, 68, 92, 130,

Index

143, 151, 156–164, 166–170, 238, 295, 307
digitalization, 83, 98
digitization, 45
dilution, 182
dimension, 200
direction, 181
disability, 267
disadvantage, 46, 333
disaster, 67, 328–331
disclosure, 215
discomfort, 26
discovery, 226
discrimination, 103, 169, 266, 267, 269, 270
disease, 50, 105, 107
displacement, 14–16, 27, 30, 33, 34, 45–49, 53, 77, 131, 132, 243–246, 250, 253, 254, 260, 294
disposal, 34, 299, 300, 304, 336
disruption, 16, 56–58, 116, 118, 132
dissemination, 58, 214, 229, 240
dissipation, 330
distance, 129, 284, 329
distribution, 3, 18, 22, 40, 53, 78, 214, 261, 272, 276, 282, 316, 324, 330
diversification, 66, 199
diversity, 14, 170, 263, 264, 266–269, 297, 299, 303, 332, 333, 337
divide, 5, 22, 24, 31, 44, 46, 61, 132, 133, 137, 160, 170, 235, 257–261, 265, 267, 268, 273, 286, 287, 289–292, 296–298, 317, 327, 335, 336
division, 282

domain, 177
dominance, 239, 289
down, 265, 271
downtime, 79, 114
drive, 1, 3, 5, 28, 51–53, 58, 68, 70, 115, 116, 130, 131, 135, 137, 141, 145, 151, 155, 158, 160, 163, 165, 173, 175, 177, 178, 183, 186, 203, 208, 219, 222, 223, 225–227, 234, 247, 269, 272, 279, 287, 289, 298, 299, 304
driver, 6, 9, 39, 207, 247, 249, 271
driving, 1, 4, 14, 15, 33, 49, 51, 55, 58, 63, 68, 70, 71, 76, 94, 97, 113, 118, 128, 130, 136, 141, 150, 153, 159, 160, 166, 173, 177, 180, 193, 200, 227, 230, 248, 269, 273, 274, 277, 285, 289, 292, 297, 298, 314, 320
drone, 323
drug, 105, 107, 216, 226
duplication, 160
duration, 214
dynamic, 3, 177, 206, 219, 227

e, 4, 45, 48, 70, 94, 192, 199, 215, 248, 266, 284, 287, 288
Earth, 329
earth, 127, 275
ease, 135, 136, 138, 142, 159, 166, 214, 272
eco, 34, 304
economic, 1–6, 9, 10, 14–16, 18, 20, 23, 24, 28, 31, 33, 36–41, 44–47, 49, 51–54, 56, 58,

63, 65–68, 70–72, 75–78, 81, 83, 85, 86, 88–94, 97–100, 113, 116–118, 122, 126–133, 135, 138, 141, 142, 145, 150, 153, 155–160, 162, 163, 173, 175, 177, 178, 180, 183–186, 188, 190, 193, 198, 201, 207, 210, 215, 222, 223, 225, 227, 228, 230, 234–237, 240, 248, 249, 251, 252, 255, 259–261, 264–266, 270–272, 277, 279, 280, 286, 288–291, 293, 296–299, 303–305, 308, 313, 314, 317, 335, 338

economy, 3, 14, 22, 44–49, 56, 68, 70, 77, 82, 86, 98, 130, 131, 159, 170, 178, 180, 185, 205, 206, 260–262, 268, 273, 276, 277, 279, 283, 288–292, 295–297, 300, 304, 305, 336–338

ecosystem, 58, 93, 94, 125, 151, 155, 180, 183, 185, 186, 188, 189, 193, 195, 197–199, 205, 209, 225, 230, 299, 332–334

edge, 21, 95, 101, 116, 214

editing, 130

education, 4, 5, 14–16, 22–24, 27, 29, 30, 32, 33, 46, 53, 58, 67, 93, 94, 131, 137, 138, 147–150, 185, 186, 197, 209, 210, 256, 257, 260, 262–265, 268, 269, 294, 299, 300, 304, 337

effect, 77, 183, 239, 243–246, 253, 254, 266, 288

effectiveness, 39, 78, 107, 148, 189, 223, 237, 238, 240, 257, 286, 300, 332, 335

efficacy, 105

efficiency, 4, 6, 9, 20, 21, 23, 34, 39, 42, 45, 48–50, 52, 53, 56, 68, 71, 78–81, 84, 93, 95, 106, 112, 113, 118, 129, 142, 160, 163, 204, 243, 245, 246, 249, 252, 272, 279, 282–285, 287, 289, 294, 308, 311, 314, 317, 336

effort, 171, 259, 279

electric, 39

electricity, 24, 29, 38–41, 55, 56, 67, 78, 110, 111, 137, 183, 290, 305

electronic, 34, 104, 107, 149, 273, 284, 304, 336

electronics, 21, 92, 104, 106, 107, 274, 276, 282

element, 305

email, 42, 272

emergence, 3, 4, 14, 15, 17, 93, 94, 223, 275, 282, 284

emergency, 130, 329

emission, 305, 306

empathy, 253

emphasis, 34, 92, 93, 189, 209, 257

employability, 15, 23, 24, 268

employee, 27, 205, 215

employment, 3, 15, 17, 39, 46–48, 54, 61, 67, 70, 77, 128, 184–186, 228, 241, 245, 246, 248, 249, 253–259, 263, 265, 266, 269

empowerment, 25, 201, 259, 265, 266
enabler, 10, 72, 83, 265
encompass, 109, 112, 123, 190, 265, 291, 305, 307, 309, 321
encryption, 27, 169, 286, 295
end, 50, 336
endeavor, 269
endurance, 114
energy, 4, 15, 25, 34, 38, 40, 41, 47–51, 54–56, 66–68, 71, 77–82, 89, 104–112, 130, 132, 142, 201, 204, 209, 229, 294, 296, 300, 303, 305–312, 316, 317, 330–332, 335, 336
enforcement, 31, 193, 215, 334
engagement, 27, 55, 173, 304, 333
engine, 199, 272
engineering, 101, 103, 105, 247, 320, 330
enhancement, 53
enterprise, 66, 201, 328
entertainment, 40
entity, 214
entrepreneur, 181, 198
entrepreneurship, 14, 16, 20, 45, 58, 60, 67, 94, 128, 133, 151, 159, 173–177, 180, 183–186, 188–206, 208, 210, 227, 238, 246, 248, 256, 257, 260, 261, 266–268, 270, 273, 291, 299, 304, 308
entry, 45, 46, 59, 184, 208, 217, 218, 223, 248, 253, 290
environment, 9, 13, 27, 30–32, 53, 54, 58, 60, 61, 67, 68, 71, 80, 94, 95, 107, 116, 123, 132, 138, 139, 143, 145, 151, 155, 173, 179, 184, 186, 187, 193, 198, 207–210, 217, 219, 223, 235, 237, 240, 248, 254, 256, 257, 263, 273, 279, 287, 297, 303, 306, 332
equality, 261, 266, 337
equipment, 26, 39, 78, 263
equity, 129, 170, 180, 182, 235, 296, 297, 300, 303
era, 32, 253, 254, 257, 264, 267, 289, 298, 321
Erik Brynjolfsson, 17
erosion, 297
essentiality, 218
establishment, 26, 274
Estonia, 93, 94
ethnicity, 324
Europe, 36
evacuation, 329
evaluation, 33, 223, 270, 300, 333
evaporation, 307
event, 187
evidence, 334
evolution, 44
example, 2–4, 14, 15, 17, 19, 21, 23, 34, 39, 42, 45, 47–50, 52, 55, 56, 69, 102, 113, 114, 125, 129, 131, 135, 142, 149, 159, 180, 185, 197, 198, 204, 216, 245–247, 253, 259, 264–266, 274, 275, 279, 282, 285, 287–289, 293, 303, 304, 307, 316, 323, 329, 330, 332–336
excellence, 132
exchange, 28, 40, 44, 71, 82, 133,

161, 163, 180, 208, 220, 228, 230, 234, 240, 272, 276, 284, 288, 296, 298
exclusion, 170
exclusivity, 216
exercise, 200
exit, 182
expansion, 3, 39, 52, 70, 77, 137, 183, 276
experience, 3, 38, 44, 48, 49, 52, 81, 148, 150, 181, 198, 199, 260, 289
experimentation, 151, 208
expertise, 3, 95, 97, 139, 142, 151, 155, 160, 181–184, 187, 209, 226–228, 230, 231, 234, 240, 246, 247, 264, 269, 276, 283, 295, 298
exploitation, 261, 275, 334
export, 66, 272, 282, 305
expression, 214
extent, 265
extraction, 279, 300, 336

fabrication, 104, 105, 114
face, 15, 22, 26, 31, 34, 48, 58, 76, 95, 137, 138, 142, 154, 163, 172, 176, 179–181, 183, 184, 188, 194, 205, 209, 250, 264, 265, 277, 279, 290, 297, 303, 304, 320
facilitation, 20, 22
factor, 29, 30, 39, 79, 93, 135, 136, 194, 260, 286
failure, 125, 182, 330
fairness, 27, 133, 235, 269, 300, 337
familiarity, 142
family, 136

farming, 102, 229, 305, 307, 320
fashion, 247
fear, 26, 30, 136, 139
feed, 142, 307
feedback, 11, 116, 137
fertility, 307
field, 15, 43, 101, 104–106, 203, 206, 218, 219, 228, 231, 239, 247, 262, 266, 299, 319
fight, 334
file, 192
filing, 213
film, 238
filtration, 306
finance, 10, 20, 48, 50, 93, 118, 141, 204, 208, 247
financing, 30, 32, 67, 76, 80, 81, 136, 181, 279
finding, 48, 193
fit, 189
fixing, 239
flexibility, 114, 116, 180, 182
flow, 50, 68, 75, 82, 227
focus, 4, 18, 42, 47–49, 52, 93, 94, 114, 128, 148, 149, 182, 186, 187, 189, 193, 195, 198, 205, 209, 214, 246, 253, 254, 257, 268, 282, 297, 299, 313, 324, 328
food, 23, 101, 229, 248, 318–320
footprint, 54, 80, 111, 236, 316, 335
force, 14, 209, 267, 270, 274, 301
forecasting, 127
forefront, 170, 177
forest, 323
form, 38, 75, 98, 116, 190, 214, 238
formation, 239
formulation, 32

Index 351

fossil, 39, 40, 47, 51, 54, 67, 107, 108, 111, 303, 305
foster, 27, 33, 53, 58, 60, 122, 131, 142, 145, 170–172, 184, 185, 193, 195, 198, 208, 210, 222, 223, 228, 232, 234, 237, 239, 248, 256, 261, 263–265, 269, 295
foundation, 10, 68, 83, 89, 147, 148
fragmentation, 4, 77, 289, 332
framework, 63, 66, 75, 82, 141, 151, 167, 194, 208, 235, 313
fraud, 50
freedom, 114, 170
Freelancer, 289
freight, 53, 76
fuel, 9, 14, 51, 54, 105, 106, 108, 111, 182, 305
function, 2, 29, 40
functioning, 63, 68, 75, 78, 82, 328, 330
fund, 261
funding, 16, 94, 97, 128, 132, 139, 140, 180, 181, 183, 186, 187, 189, 193, 194, 197, 204, 205, 209, 210, 228, 233
future, 14, 16, 18, 33, 34, 48–51, 54, 56, 91, 107, 108, 112, 130, 173, 176, 181, 203, 206, 209, 229, 237, 238, 246, 249, 259, 267, 270, 291, 297, 301, 303, 308, 312, 313, 317, 318, 320, 328, 331, 335, 338

gain, 94, 116, 122, 125, 148, 164, 177, 215, 228, 231, 265, 272, 286, 289
game, 180
gap, 20, 22, 27, 31, 53, 131, 132, 140, 164, 172, 180, 220, 228, 248, 252, 257, 260, 263, 264, 268, 273, 290, 292, 296, 297, 324, 337
gas, 56, 78, 108, 111, 303, 305, 306, 308
gender, 33, 264–268, 324, 337
Gene, 101
gene, 130
generation, 40, 45, 47, 51, 54, 56, 67, 78, 108, 110, 111, 184, 266, 305, 330, 335
genetic, 101–103, 320, 332–334
genomic, 333
Germany, 55, 56
gig, 289
glimpse, 122
globalization, 22, 215, 255, 274–277, 288, 289, 292–300
globe, 42, 159
goal, 151, 170
good, 235
governance, 31, 94, 97, 289
government, 41, 47, 58, 60, 81, 93, 94, 97, 132, 140–145, 151, 166, 171, 173, 185, 193, 197, 198, 204, 207–210, 220, 222, 254, 256, 257, 316, 338
Grace Hopper, 41
grant, 213
graphene, 106
Great Britain, 36
greenhouse, 54, 56, 108, 111, 303, 305, 306, 308

grid, 25, 40, 51, 54–56, 80, 111, 307, 316, 330, 331
groundbreaking, 93, 177, 180, 183
groundwork, 38
group, 189
growth, 1–6, 9, 14–16, 18, 22, 28, 38–41, 44, 45, 49, 51–54, 56, 58, 65–68, 70, 71, 75–78, 81, 86, 89, 91–94, 97–101, 105, 113, 116, 118, 122, 126, 128, 130–133, 135, 138, 141, 145, 150, 153, 157, 159, 160, 162, 173, 175, 177, 178, 180–188, 190, 193, 194, 197, 198, 207, 210, 222, 223, 225, 227, 230, 240, 248–251, 257, 259, 262, 264, 269–272, 274, 277, 279, 285–287, 289–291, 298, 303, 304, 308, 314, 329, 335
guidance, 141, 142, 181–183

habitat, 77, 332–335
hacking, 286
hamper, 31, 184
hand, 3, 15, 18, 86, 95, 135, 183, 187, 253, 254, 260, 265, 304, 330
handling, 27, 141, 282, 298
hardware, 139, 140, 198
harm, 34, 239, 334
harmonization, 20
hazard, 330
head, 215
health, 102, 103, 107, 136, 216, 306, 307, 333

healthcare, 4, 10, 23, 24, 28, 45, 47, 49, 50, 53, 67, 93, 101, 103, 118, 129, 130, 141, 149, 150, 166, 204, 207, 228, 247, 324
heat, 111
heating, 40, 111, 311
help, 26, 31, 33, 42, 49, 76, 77, 118, 139–141, 160, 169–171, 181, 187, 205, 206, 220, 235, 247, 254, 261, 263, 265–267, 269, 286, 297–300, 303, 304, 307, 316, 329, 333, 334
heritage, 214
hiring, 26, 266, 269
history, 6, 38
home, 201
horizon, 182
house, 226, 283
household, 38, 279
housing, 274
how, 9, 17–19, 23, 32, 33, 38, 56, 68, 71, 81, 93, 95, 126, 149, 173, 175, 176, 185, 203, 251, 252, 259, 262, 271, 279, 280, 324–326, 328, 333
hub, 70, 81, 92, 276
human, 3, 14, 24, 34, 38, 47, 49, 51, 52, 93, 103, 107, 114, 116, 195, 213, 235, 243, 245, 246, 253, 275, 306, 328
hydropower, 54, 112, 305
hype, 199

idea, 262
identification, 21, 102, 115, 231, 331, 333, 334

Index 353

identify, 16, 21, 26, 33, 52, 114, 148, 164, 171, 174, 193, 214, 230, 269, 300, 323, 330, 333, 334
identity, 122, 214, 286
image, 199
imagery, 323, 333
imaging, 323
impact, 1, 3, 6, 9, 14, 16–18, 21, 27, 31, 33, 34, 36, 38, 39, 41, 42, 44, 46, 48, 49, 53, 54, 66, 69, 71, 76–79, 98, 102, 103, 110, 112, 130, 136, 147, 149, 157, 159, 160, 167, 172, 177, 180, 185, 186, 188, 189, 192, 193, 197, 198, 200–204, 217, 227, 229, 232, 234–236, 238–241, 249–254, 260, 274, 275, 279, 280, 284–287, 296, 297, 300, 303, 305, 313, 314, 330, 336
implementation, 7, 21, 26, 31, 50, 56, 89, 96, 97, 99, 115, 116, 123, 138–140, 142, 164, 166, 171, 173, 217, 224, 267, 320, 328, 331, 335, 337
implication, 48
importance, 17, 22, 54, 66, 71, 77, 81, 88, 94, 100, 122, 125, 151, 166, 173, 227, 230, 235, 248, 249, 259, 262, 308, 321
improvement, 229, 300
incentive, 307
incident, 125
include, 5, 16, 24, 32, 37, 43, 55, 67, 70, 71, 80–82, 87, 96, 115, 116, 119, 120, 123, 128, 139, 142–144, 148, 152, 154, 157, 158, 161–163, 171, 172, 176, 188, 199, 208, 215, 220, 224, 225, 229, 233, 234, 238, 265, 284, 287, 298–300, 309, 318, 321, 329–332, 337
inclusion, 20, 22, 23, 25, 50, 75, 118–122, 131, 133, 159, 165, 185, 263, 269, 279, 290, 291, 293, 295, 297, 298, 300, 304, 325, 327, 336, 338
inclusiveness, 256, 332
inclusivity, 27, 55, 67, 68, 203, 223, 264, 267, 269, 299, 324, 328
income, 2, 3, 18–20, 23, 24, 33, 45, 67, 132, 136, 216, 227, 248, 254, 257, 259–261, 265, 268, 304, 324
inconvenience, 136
incorporation, 307
increase, 3, 23, 48, 69, 76, 128, 132, 136, 246, 254, 260, 263, 296
incubator, 189
independence, 54
India, 69–71, 197, 198, 274, 320
individual, 57, 135, 142, 147, 258, 295, 298
industrialization, 38, 81
industry, 13, 17, 26, 28, 38, 40, 43, 44, 47, 51, 80, 93, 94, 101, 102, 115, 132, 133, 139, 141, 143, 155, 166, 180, 181, 183, 192, 200, 205,

206, 208, 216, 220,
226–228, 236, 238–240,
245, 253, 262, 263, 266,
269, 274, 276, 282, 293,
299, 309, 331, 336, 338
inequality, 3, 18–20, 33, 46, 53, 67,
132, 241, 254, 259–261,
265, 268, 269, 294, 299,
324
influence, 14, 16, 17, 30, 58, 65,
135–138, 142, 147, 148,
156, 205, 260
information, 4, 9, 22–24, 27, 28, 31,
32, 39–41, 43–46, 60, 63,
68, 71, 75, 82, 98, 103,
122, 123, 125, 131, 135,
140, 147, 169, 209, 215,
235, 260, 262, 273, 284,
285, 288, 304, 321, 323,
329, 333, 334
infrastructure, 22, 24, 26, 29, 31, 32,
34, 40, 45, 51, 55, 56, 63,
65–95, 97, 100, 112, 122,
129, 132, 133, 136–139,
141, 160, 166, 183, 185,
186, 195, 197, 204, 235,
257, 259, 267–269, 271,
273, 279, 285, 286, 290,
291, 296, 298, 307, 323,
328–332
infringement, 27, 193, 214, 215
ingenuity, 213
initiative, 197, 198, 226, 256, 259,
266
innovation, 1–3, 6, 9–14, 16, 27, 30,
31, 39, 46, 47, 49, 51–53,
55, 58–61, 65, 68, 71, 75,
78, 80, 81, 83, 84, 89, 91,
93–95, 97, 115, 118, 122,

128, 130–133, 139, 141,
142, 145, 147, 151,
153–155, 160, 173, 175,
177, 178, 180, 183–186,
190, 193, 197, 198, 200,
203, 205, 207–210,
213–220, 222–228,
230–240, 246, 247, 256,
257, 260, 261, 266, 268,
270, 272–274, 276, 279,
283, 285, 289, 291,
293–295, 298–300, 304,
307, 308, 312, 314, 331,
336, 338
input, 2, 78, 249, 308
inquiry, 228
insecticide, 102
inspiration, 200, 230, 269
instability, 31
installation, 47, 51, 54
instance, 18, 21, 47, 51, 102, 130,
131, 142, 149, 245–248,
268, 282, 285, 304, 305,
307, 330, 336
insulation, 40
insulin, 101
integration, 20, 48, 51, 56, 65–67,
76–80, 106, 138, 140, 147,
148, 150, 198, 220, 227,
257, 264, 274, 276, 281,
283, 287, 291, 292, 295,
303, 307, 316
integrity, 122
intelligence, 2, 16, 21, 27, 33, 51, 93,
112, 116, 118, 209, 243,
253, 254, 264, 272, 337
intensity, 329
interaction, 228, 253
interconnectedness, 279

Index

interdependence, 275
interest, 97, 142, 216
intermittency, 51, 56, 112
internet, 22–24, 29, 31, 43–46, 69, 70, 82, 92, 94, 97, 98, 127, 132, 137, 172, 183, 184, 199, 214, 257, 260–262, 265, 267, 268, 273, 284, 286, 288, 290, 296, 298, 329, 337
interoperability, 26, 28, 50, 118, 122, 136, 141, 204, 217, 293, 332
interplay, 3
intersection, 173, 174, 176, 177, 254, 324
intervention, 209
introduction, 2, 3, 15, 44, 56, 101, 198, 246, 247, 279, 333
invention, 39, 193
inventory, 44
investment, 5, 13, 16, 22, 24, 26, 30, 51, 55, 58, 65–67, 69, 71, 75, 76, 78–81, 94, 95, 97, 132, 133, 140, 142, 180, 182, 184, 185, 193, 209, 227, 238, 264, 271–274, 286, 305, 307, 308, 335
investor, 187
invoicing, 273
involvement, 240, 333, 334
iron, 36
irrigation, 50, 307, 320
isolation, 330
issue, 18, 19, 23, 32, 34, 159, 218, 298, 329
it, 1, 3–5, 8, 9, 12, 14, 15, 18, 22, 23, 25, 26, 28–30, 32–34, 37, 38, 41, 43–46, 48, 52, 53, 56, 66, 67, 75, 76, 78, 86, 101, 104, 105, 119, 124, 135, 136, 142, 144, 156, 157, 160, 164, 167, 168, 170–172, 174, 182, 184, 198, 200, 202, 203, 207, 209, 214, 215, 217, 219, 226, 230, 232, 234, 235, 243, 245, 246, 248, 250, 251, 254, 256, 259, 261, 265–269, 271, 273–275, 277, 283, 284, 291, 295–299, 304, 315, 318, 320, 324, 327, 332, 335, 337

Jack Ma, 199
Japan, 276
job, 3, 6, 14–18, 23, 26, 27, 30, 33, 34, 45–49, 51–53, 67, 76, 92, 94, 131, 139, 148, 180, 184, 227, 243–248, 250, 252–257, 260, 262–265, 291, 294, 304, 305, 308, 314
John von Neumann, 41
journey, 179
jurisdiction, 214
justice, 235

keeping, 119, 215, 219, 226, 273
Kenya, 159, 160, 164, 185, 279
Kenya, 159, 164
Kleiner Perkins, 182
know, 95
knowledge, 2, 3, 10, 23, 25, 29, 31, 32, 43, 58, 60, 95, 131, 132, 135, 139, 142, 147–151, 160, 163, 177,

181, 184, 185, 208, 214, 215, 220, 224, 226–231, 234, 238–240, 247, 261, 262, 264, 265, 269, 272, 276, 288–290, 293, 296–298, 333, 334, 336

labor, 3, 4, 14–16, 33, 35, 45, 46, 48, 49, 67, 76, 114, 118, 209, 245, 252–257, 261, 274–276, 282, 289, 296, 297, 300
lack, 23, 26, 29–32, 129, 138, 139, 141, 142, 181, 183, 227, 254, 286, 289–291, 307
land, 318, 329, 333
landscape, 6, 20, 41, 47, 48, 133, 183, 203, 205, 214, 217, 240, 246, 249, 255, 262, 268, 279, 287, 289, 324, 329
language, 52, 233, 265, 287
Larry Page, 199
launch, 122, 128
law, 238, 334
layer, 114, 330
lead, 1, 3, 6, 14–16, 18, 22, 26, 27, 33, 45, 46, 52, 53, 60, 79, 91, 95, 132, 169, 182, 207, 209, 248–250, 254, 260, 289, 291, 297, 304, 332, 333
leader, 60, 93, 209
leadership, 200, 265, 266
leapfrogging, 186, 279
learning, 4, 15–17, 26–28, 33, 45, 48, 52, 58, 114, 116, 118, 139, 148, 150, 166, 205, 230, 248, 253, 254, 256, 262–265, 270, 272, 323
leasing, 140
led, 15, 18, 21, 42, 45, 46, 48, 51, 69, 101, 105, 106, 214, 228, 245, 252, 253, 260, 274, 276, 296, 303, 330
ledger, 119, 204
legacy, 26, 140
level, 2, 34, 105, 114, 136, 147, 218, 219, 239, 257, 266, 283, 299
leverage, 21, 44, 46, 65, 67, 70, 93, 94, 116, 130, 137, 147, 150, 160, 173, 183, 204, 206, 248–250, 262, 276, 286, 288, 290, 317, 335
liability, 34
license, 238
licensing, 227, 240
life, 5, 25, 32, 40, 55, 67, 93, 97, 102, 181, 184, 214, 216, 228, 262, 317, 336
lifeblood, 78
lifecycle, 299, 304
lifespan, 336
light, 105, 122
lighting, 39, 40, 311, 316
limitation, 291
line, 14, 39, 245, 253, 304, 335, 338
literacy, 5, 24, 29, 32, 46, 94, 100, 137, 139, 147, 150, 172, 173, 235, 236, 248, 257, 259, 261, 262, 264, 265, 267, 268, 273, 290, 291, 296, 298, 299, 337
literature, 238
litigation, 213
livability, 314

Index 357

livestock, 50, 102
living, 1, 24, 38, 40, 101, 313, 332
load, 331
loading, 76
localization, 298
location, 170, 248, 268, 274, 324, 334, 336
logging, 323
look, 110, 157, 320
loop, 11, 137
loss, 22, 26, 182, 307, 321, 332
lower, 23, 54, 66, 76, 104, 114, 255, 274–276, 296, 305, 312
loyalty, 137, 215
lure, 276

machine, 16, 35, 45, 52, 114, 116, 118, 253, 254, 272, 323
machinery, 23, 39, 78, 79
magnitude, 329
maintenance, 17, 47, 52, 54, 65, 76, 80, 114, 136, 253, 330
making, 2, 4, 21, 27, 30, 33, 42, 43, 45, 48, 51, 52, 66, 104, 106, 114, 116, 133, 136, 139, 142, 167, 170, 184, 199, 204, 213, 219, 235, 236, 238, 243, 250, 285, 289, 303, 304, 307, 314, 316, 321, 323, 334, 335
management, 4, 21, 22, 27, 42, 44, 50, 51, 53, 67, 70, 79, 127, 130, 139, 181, 182, 192, 204, 223, 229, 236, 247, 262, 272, 275, 277, 305, 308, 313, 317, 320–323, 328–331
manipulation, 101, 103, 104

manner, 119, 170, 236, 288, 295, 299, 338
manual, 2, 3, 35, 114, 245, 253, 284
manufacture, 283, 336
manufacturing, 4, 10, 14, 21, 23, 34, 35, 39, 42, 48, 49, 51, 52, 54, 77–79, 81, 112–116, 142, 215, 243, 245–247, 253, 264, 274, 276, 282, 300
mapping, 102, 330
marginalization, 22, 170
market, 3, 4, 14–17, 21, 23, 30, 43, 46, 48, 49, 66, 76–80, 114, 118, 174, 177, 180, 181, 184, 189, 193, 199, 200, 205–207, 209, 215–219, 227, 239, 243, 246, 248, 250, 252–254, 256, 257, 261–264, 272, 274–276, 283, 285–289
marketing, 22, 30, 42, 44, 45, 118, 199, 202, 204, 214, 215, 247, 248, 287
marketplace, 21, 22, 193, 214, 215, 227, 271, 274, 287, 289
mass, 38, 39
material, 114, 300
matter, 103, 104, 107
means, 182, 213, 262, 266, 284, 329
mechanism, 122
mechanization, 38
media, 45, 136, 180, 247, 272, 288
medicine, 50, 101, 103–105, 107, 130
medium, 45, 70, 140
member, 293
memory, 105
MENA, 266

mentoring, 208
mentorship, 181, 186, 187, 189, 205, 266, 268
messaging, 42, 272
micro, 20
microfinance, 19, 23, 25
migration, 140
milling, 114
million, 125, 164
mind, 126, 213
mindset, 30, 185, 262
mining, 36, 78, 128
minority, 264–267
misappropriation, 215
misidentification, 169
mismatch, 263
mission, 200
misuse, 103, 169, 218, 235, 295
mitigation, 137, 275
ML, 51
Mobile, 23
mobile, 4, 19, 20, 25, 40, 69, 131, 159, 160, 164, 185, 202, 248, 271, 279, 288, 316, 329
mobility, 49, 316, 317
model, 45, 97, 114, 201, 282, 317
modeling, 329
modification, 101
momentum, 47
money, 25, 159, 164, 185, 279
monitor, 50, 205, 257, 316, 321, 329, 330, 333
monitoring, 17, 33, 50, 80, 106, 115, 193, 300, 321–323, 328–331, 333–335
monopoly, 239
month, 189
mortar, 159

motor, 39
move, 192, 320
movement, 21, 66, 75–78, 86, 272
multi, 23, 261, 265, 286, 288, 300, 328
multimedia, 98
music, 238

nanofabrication, 107
nanoscale, 103–107
nanotechnology, 104–107, 294
nation, 93
nature, 15, 47, 50, 121, 122, 161, 219, 255, 260, 265, 289
navigation, 127, 130
need, 6, 12, 14, 15, 22, 24, 25, 27, 28, 30, 34, 45, 49, 50, 55–58, 60, 68, 70, 78, 80, 81, 102, 103, 107, 111, 114, 118, 125, 132, 133, 139, 144, 148, 150, 163, 182, 185, 203–205, 209, 210, 219, 222–224, 226, 229, 233, 239, 244–248, 250, 253, 262–264, 267, 273, 275, 276, 284, 285, 288, 289, 291, 294, 304, 307, 310, 313, 315, 317, 322, 331, 334
neighboring, 66
network, 29, 44, 75, 77, 93, 94, 97, 119, 139, 159, 181, 189, 217, 272, 288
networking, 60, 185, 187, 266, 268, 272, 274, 288
New York City, 274
North America, 36
Northern Europe, 93
novelty, 130

Index

number, 54, 214, 238
nurture, 93, 186

object, 114
off, 25, 182, 228
offering, 27, 103, 105, 111, 139, 140, 150, 186, 298
office, 42
offshoring, 70, 255
oil, 78
one, 3, 18, 20, 39, 92, 94, 125, 156, 183, 189, 199, 213, 214, 218, 228, 253, 260, 282, 294, 320
online, 16, 17, 24, 44, 45, 93, 139, 147, 148, 172, 173, 247, 248, 254, 264, 272, 284, 285, 287, 288, 291, 333
operating, 53
operation, 39, 66, 328
operator, 159
opportunity, 110, 184, 187, 266, 269
optimization, 118
order, 58, 145, 170, 237
organ, 105
organization, 138, 201, 202, 224, 226, 245
origin, 273
other, 3, 14, 15, 25, 30, 36, 43, 45, 67, 77, 78, 82, 95, 135, 155, 160, 161, 165, 170, 181, 183, 186, 187, 238, 243, 253, 254, 260, 261, 265, 267, 269, 298, 304, 317, 330, 337
output, 1, 2, 6, 23, 39, 48, 52, 68, 76, 79, 249, 308
outsourcing, 70, 274, 283, 296
overhead, 285

overreliance, 209
oversight, 34
owner, 215
ownership, 3, 50, 180, 182, 234, 273

pace, 18, 22, 30, 46, 131, 132, 184, 219, 231, 234, 243, 248, 254, 257, 273, 333, 335
paperwork, 284, 285
paradox, 6
parking, 316
part, 27, 209, 337
participation, 22, 23, 289, 291, 304
partnership, 159, 182
past, 25, 69, 77, 276, 282
patent, 192, 193, 213, 216, 238
path, 14, 112
patient, 28, 47, 49, 50, 53, 101, 118, 149
pattern, 115
payment, 23, 25, 140, 159, 248, 266, 271, 279, 287
peace, 126
peer, 25, 136, 204
penetration, 69, 94
people, 23, 45, 46, 68, 75, 78, 86, 169, 170, 217, 303, 313
perception, 51, 135
performance, 52, 80, 104, 106, 116
period, 35, 38, 213, 216, 238
permission, 213, 216
pest, 101, 307
pesticide, 102
pharmaceutical, 216, 226–228
phase, 77
phenomenon, 243, 324
phone, 4, 69
pilot, 26
piracy, 214

pitch, 187
pivot, 182
place, 4, 50, 61, 66, 140, 182, 300
plan, 139, 330
planet, 303, 305
planning, 26, 45, 53, 65, 67, 71, 78, 141, 307, 313, 316
plant, 102, 103, 307
platform, 25, 95, 159, 199, 219, 287–289, 316
player, 77
playing, 43, 101, 218, 219, 239, 266, 299
poaching, 334
polarization, 48, 49, 250
policy, 5, 46, 47, 51, 56, 59, 60, 67, 80, 81, 91, 94, 143, 185, 186, 193, 205, 208, 217–219, 237, 239, 240, 255, 256, 265, 267, 293, 308, 336
policymaking, 236, 237
pollination, 131, 228, 231, 333
pollutant, 306
pollution, 40, 108, 229, 236, 300, 303–307, 335
pool, 160, 163, 184, 231, 269, 289
pooling, 95, 160, 163, 184
popularity, 111
population, 50, 67, 148, 159, 185, 279, 318, 330, 333
port, 76
portfolio, 189, 199
portion, 159, 311
position, 193, 206, 215, 264
positioning, 130, 199
potential, 3–9, 14, 16, 18–20, 23–26, 28–31, 33, 41, 44–46, 48–55, 61, 68, 70, 78, 80, 94, 100, 103–105, 107, 108, 110, 112, 115, 117–119, 121, 122, 128, 130, 131, 133, 136, 140, 141, 146, 148, 149, 156, 158–160, 163, 165–171, 174, 177, 179–186, 189, 193, 199–201, 203, 204, 207, 210, 219, 227, 230, 236, 237, 239, 248–254, 259–261, 264–267, 272–274, 277, 279, 284, 287, 289–291, 294–296, 304, 305, 307, 308, 312, 314–317, 321–324, 329, 330, 332–335, 337
poverty, 4, 6, 23–25, 67, 184, 201, 207, 304
power, 20, 25, 39, 40, 44, 47, 51, 54, 55, 58, 78, 81, 104, 122, 130, 131, 133, 160, 177, 180, 185, 186, 198, 200, 203, 217, 218, 227, 232, 237, 239, 249, 252, 260, 279, 288, 289, 299, 303, 305, 309, 330, 331, 338
powerhouse, 77, 173
practice, 149
precision, 21, 42, 52, 113, 114, 245
preference, 30
presence, 76, 191, 247, 275, 284, 285
preservation, 295, 332–334
pressure, 136, 182
price, 79, 239
pricing, 40, 218
printing, 48, 114, 247
privacy, 22, 27, 28, 32, 44, 46, 48–50, 53, 60, 61, 118, 122, 130, 131, 133, 136,

Index

138, 140, 169, 170, 203, 205, 235, 236, 256, 273, 285, 286, 289, 295, 297, 298, 317, 323, 334, 337
problem, 9, 15, 51, 52, 114, 147, 263, 264
procedure, 101
process, 2, 10, 25–28, 33, 41, 42, 114, 135, 139, 142, 156, 160, 163, 167, 168, 170, 193, 213, 215, 226, 227, 236, 265, 270, 288, 289, 321
processing, 39, 42, 44, 52, 63, 82, 98, 169, 329
product, 48–50, 52, 114, 115, 181, 189, 198, 199, 202, 205, 217, 238, 273, 279, 287, 300, 336
production, 2, 4, 21, 23, 30, 38–40, 48, 52, 76, 79, 81, 101, 107, 112, 114–116, 216, 236, 245, 246, 272, 275, 276, 279, 282, 283, 289, 300, 304, 306, 336
productivity, 1, 2, 6–9, 15, 18, 21, 23, 25, 32, 39, 40, 42, 45–50, 52, 53, 66–71, 76, 78, 79, 81, 84, 89, 91, 101–103, 112, 114, 115, 129, 142, 147, 180, 207, 229, 243, 246, 247, 249–252, 254, 272, 274, 283, 291, 304, 311, 318, 320
professional, 147, 150, 262, 264, 272, 284, 288, 290
proficiency, 264
profit, 171, 173, 213

profitability, 250
program, 189, 223, 266, 293, 316
programming, 247, 253, 260
progress, 2, 14, 15, 18, 90, 91, 93, 133, 187, 189, 209, 219, 240, 246, 298, 300, 303, 316
project, 97, 247
proliferation, 45, 131, 204
promise, 101, 110, 120, 324
promotion, 5, 210, 303
proof, 77, 273
property, 3, 30, 31, 46, 103, 163, 185, 190–193, 205, 208–210, 213, 215–219, 226, 227, 230, 233, 240, 277, 283
proposition, 204
prosecution, 334
prosperity, 251, 335
protection, 22, 27, 32, 34, 46, 58, 106, 140, 163, 170, 190, 192–194, 205, 208, 209, 214–216, 230, 235, 238, 239, 256, 257, 261, 273, 277, 283, 285, 289, 298, 300, 303, 323, 334, 337
prototype, 181
prototyping, 114
provision, 4, 142, 186
public, 39, 44, 50, 65, 70, 76, 77, 80, 93, 97, 133, 142, 166, 170, 186, 208, 209, 214, 216, 236, 240, 265, 304, 316, 333
purchase, 26, 44, 136, 273, 287
purpose, 44, 190
pursuit, 275
push, 128

quality, 5, 21, 24, 25, 28, 29, 32, 34, 40, 45, 47–49, 52, 55, 65, 67, 68, 93, 95, 97, 102, 112, 115, 116, 118, 129, 137, 150, 184, 215, 238, 240, 253, 260, 265, 268, 272, 273, 283, 306, 308, 317
quantity, 240
quantum, 104, 105
question, 32

race, 33, 324
rainforest, 323
range, 17, 40, 44, 45, 63, 71, 75, 82, 89, 93, 101, 106, 112, 116, 123, 127, 133, 159, 186, 199, 205, 224, 243, 265, 269, 287, 305, 309, 321
rate, 42, 182, 318
reach, 22, 24, 46, 70, 76, 159, 185, 193, 200, 202, 206, 227, 247, 248, 266, 271, 284–288
realm, 46, 264
reassessment, 169
recognition, 51, 115, 137, 169, 170, 200, 214, 238, 264, 304
reconstruction, 330
record, 119, 149, 189
recourse, 214
recovery, 328, 330
recruitment, 265
recycling, 34, 300, 304, 305, 336
redistribution, 33, 257, 261
reduction, 4, 23–25, 54, 116, 184, 275, 276, 282, 287, 300, 305
redundancy, 77

refinement, 300
reflectivity, 106
refrigeration, 39, 40
regeneration, 105, 336
regime, 191
region, 76, 180, 209, 266, 334
registration, 94, 185, 214, 215
regulation, 138, 239, 289
reinforcement, 116
relationship, 3, 6, 9, 14, 18, 68, 173, 174, 250, 259–261
release, 54, 105
relevance, 150
reliability, 34, 80, 159, 331
reliance, 25, 47, 107, 125, 159, 284, 333
relief, 329
remediation, 106
remote, 4, 20, 23, 24, 40, 50, 55, 129, 130, 170, 247, 265, 268, 290, 323, 328, 329, 332
renewable, 15, 34, 40, 41, 47, 51, 54–56, 67, 77, 79–81, 107–110, 112, 130, 142, 201, 209, 229, 296, 300, 303, 305, 307, 336
repair, 105
report, 293
reporting, 125, 333
representation, 265
representative, 33
reputation, 80, 189, 215, 226
research, 13, 14, 22, 30, 41, 43, 44, 51, 58, 60, 80, 93, 94, 103, 107, 122, 127, 129, 132, 151, 160, 163, 181, 185, 194, 204, 207–210, 216, 220, 223, 226–228, 230, 231, 236, 239, 269, 272,

Index 363

274, 292, 293, 298, 299, 307, 331, 335, 336, 338
reshaping, 4, 118, 280–283
residency, 94
resilience, 58, 77, 78, 80, 313, 314, 328–332
resistance, 26, 27, 30, 31, 139, 141, 142
reskill, 245, 248, 256
reskilling, 47, 48, 254, 262, 299
resource, 2, 28, 30, 40, 52, 68, 127, 129, 130, 160, 161, 163, 184, 229, 236, 275, 277, 296, 304, 308, 317, 320–323, 335, 336
respect, 233, 234
responsibility, 34, 58, 248, 292
rest, 36
restoration, 334
result, 2, 8, 15, 23, 48, 53, 60, 66, 70, 164, 231, 246, 255, 333
retail, 15, 44, 45, 48, 199
retraining, 15, 16, 27, 33
return, 140, 142, 335
reuse, 336
revenue, 181, 184, 227, 285
revolution, 44–46, 48, 291
ride, 17, 248, 288
right, 184, 193, 213, 214, 216, 235, 239, 324
rise, 39, 44–46, 48, 180, 215, 246, 248, 287, 289, 304
risk, 13, 15, 21, 22, 47, 48, 50, 137, 170, 180, 181, 183, 223, 254, 275, 291, 296, 297, 323, 330, 333
road, 29
roadmap, 264
robotic, 113

role, 1, 4, 6, 9, 14, 15, 20, 23, 30, 31, 34, 38, 40, 47, 51, 54, 58, 60, 66–68, 71, 73, 75–81, 84–86, 89, 92, 93, 97, 101, 105, 106, 111, 113, 115, 122, 124–126, 129, 130, 133, 135–138, 141–143, 145, 147, 148, 150, 151, 153, 155, 156, 160, 162, 163, 166, 174, 177, 180, 181, 183, 185, 186, 188, 189, 193–198, 200, 202, 205, 207–210, 214, 216, 217, 219, 222, 223, 227–230, 234, 235, 239, 240, 245, 246, 248, 254, 257, 262, 264, 265, 271, 273, 274, 277, 279, 281–283, 288, 289, 292, 293, 295–297, 303–305, 312, 313, 317, 318, 320, 324, 327, 328, 330, 331, 334, 336
Ronald Wayne, 198
room, 41
rotation, 307
route, 53
routine, 14, 15, 45, 48, 49, 52, 53, 250, 253–255
run, 46
runoff, 307
runway, 182

safeguard, 27, 32, 122, 235, 273, 298
safety, 33, 34, 50, 103, 107, 113, 114, 122, 172, 209, 246, 254, 293, 328
sale, 216
sanitation, 67

satellite, 126–130, 323, 329, 333
saving, 129, 216, 228
scalability, 50, 107, 122, 182, 184, 204
scale, 23, 26, 29, 32, 43, 46, 76, 79, 80, 104, 107, 111, 163, 181, 184, 185, 187, 200, 202, 266, 272, 275, 283, 284, 288, 301, 305, 307, 320, 330
scaling, 30, 183, 194, 204
scarcity, 54, 129, 184, 318
scheduling, 53
science, 93, 104, 106, 107, 205, 333
screening, 189
scrubbing, 306
scrutiny, 170
seaport, 76
search, 180, 199, 288
secrecy, 215
secret, 215
section, 1, 6, 9, 14, 18, 20, 23, 25, 29, 32, 36, 38, 41, 44–46, 49, 54, 57, 58, 63, 68, 71, 75, 78, 81, 92, 104, 107, 112, 118, 122, 126, 130, 133, 135, 138, 142, 147, 156, 160, 164, 167, 170, 173, 177, 183, 186, 193, 198, 200, 203, 207, 217, 224, 227, 230, 234, 237, 241, 243, 246, 249, 252, 254, 259, 262, 264, 267, 271, 274, 277, 279, 284, 289, 292, 295, 297, 305, 313, 318, 321, 324, 328, 332, 335
sector, 15, 39, 40, 47, 48, 54, 70, 80, 95, 97, 104, 105, 118, 119, 129, 130, 132, 133, 149, 150, 171, 183, 186, 219, 220, 227, 228, 245, 261, 269, 279, 291, 296, 299, 300, 308, 309, 318
security, 15, 21, 27, 28, 32, 50, 51, 54, 56, 79, 81, 101, 121, 122, 131, 133, 136, 138, 140, 141, 159, 170, 205, 215, 229, 247, 257, 285–287, 295, 318–320, 334
seed, 189
segmentation, 42
selection, 102, 189
self, 15, 33, 49, 106, 330
semiconductor, 282
sensing, 323, 328, 329, 332, 333
sensor, 115
separation, 105
sequencing, 333
Sergey Brin, 199
service, 39, 93, 95, 140, 159, 164, 185, 217, 243, 247, 248, 284–286, 295
set, 28, 109, 123, 193, 198, 200, 202, 215, 217, 226, 262
setting, 296
share, 17, 25, 37, 50, 55, 76, 151, 199, 237, 240, 336
sharing, 25, 28, 32, 45, 58, 60, 95, 131, 132, 160, 161, 163, 184, 208, 215, 224, 226, 227, 231, 234, 239, 240, 248, 272, 288, 289, 293, 298, 304, 316, 332
shelf, 102
shelter, 23

Index 365

shift, 2, 3, 35, 38, 182, 247, 252, 255, 276, 308
shipping, 287
shopping, 44
shortage, 184
side, 105, 303
sign, 93
signal, 181
significance, 63, 75, 78, 81, 104, 107, 125, 126, 149, 207
silicon, 104
Silicon Valley, 180, 209, 274
Silicon Valley, 180, 209
simplicity, 198
simplification, 20
Singapore, 92–94
site, 266
size, 104, 184
skepticism, 31
skill, 3, 4, 9, 15–17, 27, 44, 53, 58, 246, 252, 255, 260, 262
smartphone, 135
society, 5, 25, 28, 32, 34, 38, 45, 53, 56, 61, 66, 87, 95, 97, 133, 137, 167, 170, 172, 200, 230, 234, 236, 240, 250, 252, 254, 264, 267, 269, 291, 296, 304, 324, 327, 328, 332, 335, 337
socio, 24, 31, 290
software, 17, 26, 42, 125, 136, 139, 140, 147, 172, 192, 193, 198, 205, 214, 238, 243, 247, 253
soil, 50, 307
solution, 20, 192, 304
solving, 9, 15, 51, 52, 107, 114, 116, 263, 264
source, 26, 39, 79, 181, 266, 308

sourcing, 275, 300
South Korea, 60, 92, 94, 209, 276, 282
South Korea's, 60, 92, 209
space, 111, 126–130, 187
span, 262, 275
specialization, 282, 283
species, 332–334
speech, 51
speed, 21, 52, 70, 77, 92, 94, 97, 114, 245, 290, 333
sphere, 116, 297
spin, 105, 228
spread, 28, 29, 36, 102, 156, 160, 167, 333
spur, 11, 51, 186
stability, 105, 106, 239
stage, 180–183, 186
stakeholder, 236
start, 215, 228, 248, 266
startup, 93, 94, 182, 185, 186, 188, 192, 223, 227
status, 170, 265, 267, 293, 324, 336
steel, 79, 81
stem, 324
Steve Jobs, 198
Steve Wozniak, 198
stigma, 136
storage, 27, 47, 51, 56, 63, 78, 80, 105, 106, 235, 286, 295, 298, 307, 330
store, 41, 50, 105, 119, 285
story, 185, 199, 279
strategy, 27, 192, 335, 336
strength, 106, 114
structure, 16, 186, 333
study, 71, 77, 82, 160, 164, 173, 227, 256, 257, 276, 287, 317
subset, 116

subsidence, 329
success, 44, 56, 92–94, 159, 164, 178, 180, 183, 185, 188, 189, 193, 194, 198–200, 202, 205, 206, 209, 222, 223, 230, 240, 279, 285
suitability, 333
suite, 199
summary, 22, 32, 44, 254, 304
sun, 111
sunlight, 305
supplier, 79
supply, 4, 22, 39, 42, 47, 50, 53, 67, 70, 71, 75, 78–81, 131, 132, 185, 192, 204, 247, 272, 274–277, 290, 296, 300, 330, 331, 334
support, 15–17, 28, 30, 31, 47, 58, 60, 63, 68, 75, 80, 82, 94, 97, 132, 133, 138, 140, 142, 143, 166, 171, 172, 180, 181, 186, 187, 189, 195–197, 200, 207, 208, 220, 222, 223, 228, 230, 240, 245, 247, 248, 254, 256, 257, 261, 268, 292, 294, 296, 300, 304, 305, 329
surface, 104, 106, 329
surveillance, 32, 323, 334
sustainability, 31, 32, 39, 40, 50, 67, 77–81, 106–108, 110, 111, 113, 115, 129, 130, 133, 189, 228, 229, 235, 236, 259, 296, 299, 300, 303, 304, 312, 317, 335, 338
Sweden, 256, 257
switching, 104

system, 52, 55, 75, 76, 93, 147, 213, 226, 257, 279, 316

table, 181
Taiwan, 276, 282
taking, 13, 80, 108, 114
talent, 93, 94, 180, 181, 184, 189, 204–206, 231, 289, 308
tamper, 273
target, 21, 286, 334
targeting, 105
task, 46, 142, 308
taste, 102
tax, 94, 137, 142, 185, 208, 257, 307
taxation, 261, 287
taxi, 17
teaching, 149
team, 247
teamwork, 42, 263
tech, 92, 180, 269, 299
technique, 333
technology, 1–12, 14–32, 34, 39, 44, 48, 50, 58, 60, 63, 65, 68, 92–94, 101, 111, 114, 118–122, 125–130, 132, 133, 135–177, 180–186, 188–207, 209, 217–220, 223, 227–230, 234–241, 243, 244, 246–257, 259–277, 279–283, 286, 289–301, 303, 304, 307, 308, 313–318, 320, 321, 324–328, 331–338
telecommunication, 39, 285
telecommunications, 4, 25, 29, 66–69, 71, 74, 75, 82, 85, 89, 92, 94, 127, 164, 271, 290
telemedicine, 4, 129, 130, 149

telephone, 329
temperature, 316
term, 26, 31, 40, 45, 68, 80, 140, 181, 182, 186, 187, 223, 236, 264, 303, 313, 324, 335
testament, 180
testing, 103, 181, 236
the United States, 274, 276
theft, 122, 193, 286
therapy, 101
thickness, 104
thing, 25
thinking, 15, 150, 253, 254, 263
threat, 332, 334
tier, 93
time, 3, 4, 21, 42, 47, 52, 107, 115, 131, 182, 204, 213, 227, 262, 272, 273, 284, 285, 316, 318, 321, 323, 329–331
timeline, 182
timing, 200
Timo Boppart, 17
tissue, 105
today, 1, 20, 58, 82, 122, 173, 180, 205, 230, 246, 249, 257, 259, 262, 267, 271, 274, 276, 279, 288, 289, 292, 297, 321, 324
tool, 23, 183, 267, 327, 334
tooling, 114
topic, 249
tourism, 128
toxicity, 107
traceability, 50, 131
track, 50, 189, 272, 273, 323, 334
tracking, 300

trade, 4, 6, 20–23, 45, 66, 67, 75–78, 86, 182, 190, 213, 215, 256, 271–274, 280, 281, 283, 286, 288, 289, 296, 333, 334
trademark, 214, 215
traffic, 50
trafficking, 334
train, 202
training, 4, 16, 22, 24, 26–29, 31, 32, 53, 94, 131, 132, 137, 139–141, 143, 147–150, 170, 172, 187, 205, 215, 235, 248, 254, 256, 257, 260, 262, 263, 265, 267, 268, 286, 287, 290, 298, 299
trajectory, 1
transaction, 22, 24, 66, 68, 271, 273, 285
transfer, 25, 44, 60, 103, 129, 159, 164, 193, 207, 220, 223, 227–230, 238, 240, 285, 298, 307
transformation, 48, 49, 75, 85, 88, 93, 117, 118, 131, 135, 247, 248, 252, 293
transition, 28, 40, 47, 55, 56, 108, 112, 130, 132, 244, 245, 247, 254, 303, 305
transitioning, 40, 41
translation, 228
transmission, 39, 40, 63, 71, 78, 81, 105, 286
transparency, 32, 33, 50, 119–122, 131, 133, 141, 159, 170, 204, 236, 273, 298, 337
transplantation, 105
transport, 66

transportation, 4, 17, 29, 36, 38–40, 49, 53, 66, 68, 71, 75–78, 81, 86, 89, 183, 243, 271, 305, 306, 308, 309, 313, 316, 317, 333
travel, 24, 284, 285
treatment, 24, 53, 101, 107, 118, 169, 228, 265, 296, 305, 306, 308
triage, 118
trust, 50, 122, 125, 131, 136, 138, 204, 214, 215, 227, 238, 240, 272, 273, 285–287
turn, 14, 18, 39, 76, 246, 249
turning, 114
type, 214

U.S., 166, 209
uncertainty, 30, 79, 132, 180, 183
underemployment, 255
underrepresentation, 265
understanding, 6, 9, 17, 26, 32, 104, 125, 147, 177, 182, 193, 206, 219, 239, 265, 266, 287, 288, 324, 333
unemployment, 47, 250, 254, 255
unit, 2, 249
United States, 209
unloading, 76
up, 15, 17, 22, 31, 49, 52, 76, 80, 107, 128, 184, 194, 203, 219, 246, 253, 273, 284, 288, 323
update, 219, 254
upgrading, 58
upskill, 15, 248, 256, 264
upskilling, 15, 17, 47, 148, 205, 246, 254, 262
urbanization, 313–317

USA, 180
usability, 291
usage, 24, 50, 69, 124, 142, 172, 262, 298, 316, 335
use, 2, 16, 21, 23, 27, 28, 32, 34, 38–40, 44, 45, 48, 58, 68, 77, 79, 94, 101–103, 105, 108, 111–113, 115, 118, 127, 130–133, 135–138, 142, 147–149, 159, 166–168, 170, 172, 173, 199, 202, 203, 214, 216–218, 235, 236, 252, 262, 265, 272, 290, 293–295, 299, 300, 303–305, 307, 324, 330, 332–334, 337
usefulness, 142
user, 76, 93, 131, 169, 198, 199, 217, 289
utilization, 5, 24, 29, 68, 95, 128, 139, 166, 295

validation, 136, 181, 228
validity, 218
value, 2, 4, 9, 10, 18, 49, 52, 140, 170, 173, 193, 204, 215, 217, 272, 279–283, 288, 291
variety, 109, 305, 332
vegetation, 333
vehicle, 17, 53
venture, 30, 180–183, 200, 204
viability, 97, 313, 335
video, 42, 129, 272
view, 181, 264
vision, 115, 181, 182, 233
volatility, 79, 205, 206
volume, 104

Index 369

vulnerability, 54, 125, 330

wage, 250, 265
waiting, 76
warming, 305
warning, 328, 329, 331
wastage, 307
waste, 34, 40, 114, 116, 229, 300, 304, 305, 308, 316, 332, 335, 336
water, 54, 67, 106, 111, 305–308, 318
wave, 183
way, 10, 20, 23, 44, 98, 107, 164, 173, 180, 276, 283, 284, 287–289, 321, 335
wealth, 3, 227, 261
weapon, 218
weather, 127, 329
web, 199
website, 248
weed, 102
welfare, 102, 239, 256, 257
well, 10, 22, 24, 29, 40, 55, 63, 66–68, 76, 78–80, 101, 129, 135, 171, 184, 189, 191, 205, 210, 214, 235, 240, 246, 290, 303, 328, 332, 335, 337
whole, 14, 38, 87, 250, 254

wildlife, 333, 334
wind, 40, 47, 51, 54, 81, 111, 303, 305, 330
window, 21
word, 42
work, 16, 18, 20, 26, 33, 71, 82, 86, 91, 114, 119, 135, 149, 180, 186, 187, 205, 214, 218, 235, 246, 247, 253, 254, 263, 265, 270, 286, 293–295, 297, 328, 337
worker, 2, 16, 250, 261
workforce, 22, 47, 53, 56, 94, 116, 131, 150, 184, 195, 205, 245, 248, 252–255, 257, 263, 264, 299
working, 17, 253, 260, 261, 269, 275, 287, 289, 296, 336
workplace, 269
world, 1, 17, 20, 23, 36, 45, 46, 51, 55, 58, 81, 82, 92–94, 112, 121, 122, 131, 153, 160, 164, 173, 177, 180, 196, 199, 200, 205, 227, 228, 230, 234, 237, 246, 251, 257, 259, 262–264, 267, 271, 274, 279, 284, 287–290, 292, 295, 297, 308, 313, 320, 338